P9-CJN-574

STRONG WATER

NITRIC ACID: SOURCES, METHODS OF MANUFACTURE, AND USES

STRONG WATER

NITRIC ACID: SOURCES, METHODS OF MANUFACTURE, AND USES

Thomas H. Chilton

THE M.I.T. PRESS
Massachusetts Institute of Technology
Cambridge, Massachusetts, and London, England

First M.I.T. Press Paperback Edition, October 1970

Set in Monophoto Times New Roman
and printed by Halliday Lithography Corporation
Bound in the United States of America by The Colonial Press, Inc.

ISBN 0 262 03023 3 (hardcover)

ISBN 0 262 53015 5 (paperback)

Library of Congress catalog card number: 67-16496

To

Guy B. Taylor

who first introduced me to the complexities of the nitrogen oxides
this work is affectionately dedicated

PREFACE

"The history of the manufacture of nitric acid splendidly exemplifies the response of the inventive spirit to the interplay of political and economic forces, while the design of modern plants demonstrates the value of applying the principles of pure chemistry in a chemical engineering achievement."* The objective of this work is to fill out for the chemistry student, or for the general reader with some background in the science and an interest in technology, the truth of the statement just quoted.

While every effort has been made to see that the factual material presented is authoritative, this is not a "how-to-do-it" technical manual. For a treatment of that sort, the reader is referred to *Chemical Engineering Progress Monograph Series, No. 3*, by the same author, from which much of the matter in Chapter Four is adapted, with permission of the copyright owner, the American Institute of Chemical Engineers, hereby gratefully acknowledged.

Thanks are also rendered to E. I. du Pont de Nemours & Co., Inc., for permission to publish the above-mentioned material originally, and for supplying Figure 3.4; and for permission to quote or supplying illustrations to John Wiley & Sons, Inc., and their Interscience Publishers Division; to Columbia University Press; to Reinhold Book Division, Reinhold Publishing Corporation; to D. Van Nostrand Co., Inc.; to Oliver & Boyd, Ltd., for Gurney and Jackson; to Verlag-Chemie G.m.b.H.; to Manufacturing Chemists' Association, Inc.; to The Chemical and Industrial Corp.; to The International Nickel Company, Inc.;

*W. Fletcher and N. W. Roberts, *British Chemical Engineering*, **3**, 124 (1958).

to Encyclopaedia Britannica, Inc.; to Fritz W. Glitsch & Sons, Inc.; to The Society of Chemical Industry; to *Industrial and Engineering Chemistry*; and to *Chemical and Engineering News*.

THOMAS H. CHILTON

Hockessin, Delaware
October, 1966

CONTENTS

CHAPTER ONE

HISTORICAL; FROM THE ALCHEMISTS THROUGH WORLD WAR I

Nitric acid, a common laboratory reagent, is an important industrial commodity. It is produced in the United States at the rate of 4.6 million tons per year (1964 figures), with a market value of $350 million. Most of this goes into agricultural fertilizers, largely in the form of ammonium nitrate, a rapidly expanding outlet. The 1964 production rate of nitric acid is nearly three times as much as in 1951, and twenty-eight times that in 1939.

Uses for nitric acid, aside from small requirements for refining of precious metals, did not develop any large demand until about 100 years ago. Nitroglycerin, discovered in 1847, was introduced as an explosive by Nobel in 1860, and in the safer form of dynamite in 1867. Nitrocotton (nitrocellulose, or guncotton) was patented in 1846, though the first successful plant was not built until 1871. It was about the same time that the coal-tar dyestuffs industry began to develop, following upon Perkin's discovery of mauve in 1858; most of its raw materials were obtained by nitration. All of these required nitric acid.

Until that period, until the introduction of dynamite and of smokeless powder (based on nitrocellulose), all explosives depended upon nitrate salts, for 500 years upon potassium nitrate (niter, or saltpeter; the terms are used interchangeably), supplemented only since about 1825 with sodium nitrate (known as Chile saltpeter). Since these were, until the development of the ammonia synthesis process, the source of nitric acid, let us take a glance at the history of saltpeter.

Saltpeter and Gunpowder

The ancients apparently had no knowledge of saltpeter. There are two references in the Bible, at least in the King James Version, to "nitre." In both passages, however, in Proverbs 25:20, ". . . as vinegar upon nitre, so is he that singeth songs to an heavy heart," and in Jeremiah 2:22, "For though thou wash thee with nitre, and take thee much sope, yet thine iniquity is marked before me . . . ," it is apparent from the context that the original should have been translated as *lye* (or potash), and it is so rendered in later versions. The passages only show that Solomon was an observer of chemical phenomena as seen in the household, and that washing powders as well as soap, made from potash, were known in early times.

The first undisputed mention of saltpeter is found in the writings of the Arabian, Abd Allah, about the year 1200 A.D., who refers to it as "Chinese snow," giving rise to, or else confirming, the widely held belief that the Chinese had used saltpeter mixtures in fireworks as early as the tenth century.

The English friar, Roger Bacon (1214–1294), gave directions for making what we now call gunpowder, though he took pains to conceal in an anagram of transposed letters in his Latin text the exact proportion of an essential ingredient—a practice of which writers of patent specifications have been accused in our own day, though not by such an obvious device. The alchemist Albertus Magnus (1193–1280) describes the properties of a powder that could cause rockets to rise. Deciphered, Bacon's formula calls for 7 parts saltpeter, 5 parts charcoal, and 5 parts sulfur, which would provide a workable but not very powerful mixture, the percentages being roughly 40, 30, and 30%, compared to the present-day standard of 75% saltpeter, 15% charcoal, and 10% sulfur. You could make some, the same way that Roger Bacon did 700 years ago, by mixing the ingredients, slightly dampened with a little water, in a mortar with an apothecary's pestle—but I would not advise you to try it: too many of those who have tried

Figure 1.1 Berthold Schwartz
by Raphael Custos (1643)

"Likeness of the Reverend and Ingenious Father, of the Franciscan Order—Physician, Alchemist, and Inventor of the Liberal Art of Musket Shooting."

A translation of the verses appearing beneath the illustration—which read across the lines, not in columns—follows:

See how the fruit of time, in Nature's might,
Is by keen-witted folk oft brought to light.
The art of musketry, thus born, became
Child of Dame Nature's breath and fiery flame.

From the Berolzheimer Collection of Alchemical and Historical Reproductions, courtesy of the Chemists' Club Library, New York. Translation appears in *Industrial and Engineering Chemistry*, *26*, 802 (1934).

it have had serious accidents. Making gunpowder, even on a small scale, is a dangerous business.

It remained for a German monk, many historians seem to agree (though others say he is legendary; see Figure 1.1) to conceive the idea of confining such a rapidly burning mixture in a pressure chamber to develop enough force to propel a projectile; that is, to conceive the invention of firearms. The monk's name was Berthold Anklitzen, better known as Berthold Schwartz, or The Black, from his devotion to the "black arts," and the date of his invention is given as the year 1313. All the European kingdoms, in any event, began about this time to add firearms of one sort or another to their arsenals. But the early firearms were pretty crude and ineffective, and in the first large battle in which they were employed, the Battle of Crécy, which took place on August 26, 1346, they had no decisive effect upon the outcome. The English victory is credited to the superior effectiveness of the longbowmen over the French cavalry.

Nevertheless, the inherent capabilities of the new weapons came to be recognized, as improved designs were developed; and from that date onward we hear less and less about the bow, the spear, the sword, and more about the musket and the cannon. The military capability of every country then became dependent on a supply of saltpeter for gunpowder.

"For a very long time after the introduction of gunpowder," the article in the 1958 Edition of the Encyclopaedia Britannica tells us, "the question of an adequate supply of saltpeter was everywhere a serious problem. It was, and still is, commonly found in the surface soil in many parts of India. It is likewise present in certain parts of Spain, especially in southern Andalusia. But in England, and on the continent without the Spanish borders, no such ready-made supply existed. Hence most nations undertook to exploit the niter deposits which everywhere existed as incrustations on the walls of cellars and stables; and special agents of the crown, usually a rowdy and undesirable lot, were appointed to search out and bring these in. Later, it having been ascertained that such exudations were the result of the decomposition of organic matter, 'niter beds' were set up in practically all European countries lacking other internal sources of that commodity. These consisted of layers of decaying animal and vegetable substances, mortar from old walls, earth, sand, etc., piled to a thickness of three to four feet on floors of wood or clay. The whole placed under cover, and moistened periodically with blood or urine, was ready after the lapse of about two years for the extraction of the saltpeter which had formed over this period as the result of the interaction of the materials so mixed. In Prussia, farmers were required to bound their lands with walls of like composition, these being town down periodically for the recovery of the saltpeter elaborated within and in Sweden the countryfolk were required, until well past the middle of the nineteenth century to pay a portion of their taxes in this substance, the equivalent value in specie [coinage] not being accepted."[1]

In America the General Court of Massachusetts passed an order in 1642 requiring every plantation within the Colony to erect a niter shed. And in a much later time, efforts were being made, according to a tradition in the author's family, to revive this slow and inefficient process in the Confederate States during the American Civil War, though there is no record that any significant production was achieved.

The chemical action taking place in the "niter beds" or "niter plantations," as they were sometimes called, is the conversion, by the action of nitrifying bacteria, of the nitrogen combined as urea or resulting from the decomposition of animal or vegetable protein, in the presence of lime or potash, to nitrates. On leaching with water, with the addition of potash, the conversion to soluble potassium nitrate is completed, and this was recovered from the washings by evaporation and repeated recrystallization. The yield was low. Only about 0.3 lb of saltpeter was recovered per cubic foot of the beds every two years.

It is hard to see how the wars of the fifteenth and sixteenth centuries could be carried on with such a source of saltpeter for gunpowder. Beginning in 1626, however, saltpeter began to reach Europe from India, and local supplies diminished in importance.

[1] "Gunpowder," Vol. 11, p. 4.

In the more arid regions of India, such as the Bihar section of the plain of the Ganges, a combination of climatic and hydrologic conditions and sanitary practices of the population makes it possible to collect regularly an efflorescence on the surface of the soil rich in saltpeter. The drains from the village houses and cow-sheds receive mainly urine, and run to an open plot where they evaporate or soak into the ground; the only other material reaching this area consists of ashes from the fires used in cooking. Niter is abundantly produced only in those localities where the soil is sandy and where the water table is neither too deep nor too shallow. Liquid rises to the surface to replace liquid lost by evaporation, and soluble salts including saltpeter are deposited at or near the surface. There is or used to be a caste of people who obtained their livelihood through the collection and refining of saltpeter. In addition to the replenishable supply, deposits from abandoned village sites are worked.

The Indian saltpeter trade was for a long time a monopoly of the British East India Company, which was under an obligation to supply the British Government before other consumers. Coupled with naval power to enforce a blockade, this trade must have been a tremendous advantage to the British in any war they were engaged in during that period, though I do not recall any allusion to that aspect in my history books: Napoleon must have had to rely on niter plantations, or deposits in Spain, or on blockade running.

Two incidents connected with American history are recorded. E. I. du Pont, who had sought the counsel of Thomas Jefferson before going into the manufacture of gunpowder, in turn advised Jefferson to buy $50,000 worth of saltpeter, selling at 16 to 20 cents a pound. This was of great benefit when the War of 1812 shut off the supply from India, and saltpeter, produced from deposits of calcium nitrate found in the Mammoth Cave in Kentucky, sold for 32 to 38 cents a pound.

The second was during the Civil War. The campaigns of 1861 had depleted the gunpowder supplies of the North; and Great Britain, who controlled the nitrate trade, had recognized the Confederacy as a belligerent. Under these circumstances, late in

1861, Lammot du Pont, one of the younger members of the firm founded 60 years before by his grandfather, was asked to go to London and purchase all the saltpeter he could acquire. He succeeded in obtaining 2000 tons, and it was being loaded for shipment when the British Government imposed an embargo on it, as a result of the seizure of the Confederate commissioners Slidell and Mason from the British mail ship *Trent* by the U.S.S. *San Jacinto*. Lammot du Pont returned to Washington to report the situation, about the same time as the British issued an ultimatum demanding the release of the commissioners. Though Congress had passed a vote of thanks to the officers of the *San Jacinto*, the U.S. Government released the commissioners. They took an English ship for Europe on January 1, 1862; Lammot du Pont sailed on another ship on the same day for London. The embargo was lifted on January 18, and ships carrying £80,000 worth of saltpeter sailed for the United States on February 2. The Confederacy, according to their Chief of Ordnance, obtained about half of their saltpeter supply from limestone caverns, the remainder being brought in by blockade runners from English sources.

The production of saltpeter in India reached its peak during this period, reaching about 33,000 tons per year. It declined to about half that rate by 1913, and while stimulated during World War I to 25,000 tons per year, it has declined again, at latest account practically to the vanishing point.

A remnant of the process for producing saltpeter by soil efflorescence persists on the American continent, or did as late as 1919, when it was observed in Guatemala by engineers looking for deposits of nitrate. The product is used locally in fireworks for celebrations. The process is essentially the same as in India, the main difference being that it is carried on only in the dry season (November–May), and by women. The streets in the villages serve for drains, and sweepings from the streets are mixed with wood ashes and leached.

Modern industrial chemical laboratories are coming more and more to employ instrumental methods of analysis for control of manufacturing operations, and here is an instance where very primitive technology was employing effective, if unsophisticated, devices. The effort of reconcentration and crystallization of the saltpeter would depend on the amount of water used for leaching the salts from the sweepings from the street. When should you stop the extraction? A handy device for determining the density of a liquid, and thus inferring the concentration of dissolved material in water solutions, is the hydrometer. The hydrometer consists of a weighted glass bulb with an elongated stem, which sinks, by Archimedes' principle, to a depth where the weight of the liquid displaced equals the weight of the instrument. To determine when the water extract had become so dilute as not to justify further effort, the Guatemalans used a hen's egg as a hydrometer; when it ceased to float, they stopped adding water.

The liquor so obtained was concentrated by boiling in earthenware pots. Crystals are deposited on cooling, and are purified by recrystallization, the whites of eggs being used as a clarifying or coagulating agent—thus employing a principle of colloid chemistry often used in more scientifically controlled operations.[2]

Saltpeter, nitrate of potash, is still the principal ingredient of gunpowder, and gunpowder was the main reliance for military explosives up through the Spanish-American War. Dependence upon local workings or upon importation from India, however, was relieved first by the discovery and exploitation of deposits of sodium nitrate in South America, and then by that of potash minerals in Germany. The conversion of sodium nitrate to potassium nitrate by means of potassium chloride was introduced about 1860 and was used in the United States during the later years of the Civil War. We now take leave, for the time being, of potassium nitrate and take up the story of nitrate of soda.

Chile Saltpeter, or Nitrate of Soda

In 1809, a German by the name of Haenke, living in Bolivia, called attention to the beneficial effects of a natural nitrate as an

[2] H. S. Gale, "Saltpeter in Guatemala," *Engineering and Mining Journal*, **107**, 1025–1031 (1919).

agricultural fertilizer. Exploitation of the deposits commenced soon after, and exportation began, in a small way, in 1826, reaching 100 tons/year by 1831. The principal use was in agriculture, both in Europe and in the United States. It was not long before ways were found to use it in "black powder," for commercial explosives, if not for military. On May 19, 1857, Lammot du Pont obtained a patent for a blasting powder composition consisting of refined "Peruvian" nitrate of soda, with charcoal and sulfur, "glazed" with graphite, which soon came to dominate the field. With the introduction of the potassium chloride conversion process, utilizing the German deposits opened up in the 1850s, the demand again increased. It increased still further with the demand for nitric acid for the manufacture of nitroglycerin and nitrocellulose and dyestuff intermediates, as mentioned earlier. By 1870 production had grown to over 100,000 tons/year.

The deposits were found in parts of what were Bolivia and Peru, as well as Chile. In 1879, Chile engaged in a war with these countries, and gained control of practically all of the nitrate fields. From then for more than thirty years, Chile enjoyed a virtual monopoly on the supplies of nitrate of soda, or Chile saltpeter, as it came to be called. Production rose to nearly 1 million tons/year by 1890, and exceeded 3 million tons/year in the last year of World War I, 1918. During this period, the largest part of the revenues of the Chilean Government was derived from a tax of $12 a ton on the exportation of nitrate. The industry suffered a setback upon the emergence of synthetic ammonia, in the years following World War I, and then the economic depression in the 1930s. The nitrate producers and the European producers of synthetic nitrogen compounds attempted to work out agreements on prices and production quotas by means of a succession of what are known in Europe as cartels, but these efforts were never entirely successful. Natural economic laws appear to have finally prevailed, and production, in competition with synthetic nitrate of soda, produced from ammonia, has apparently stabilized at a level of about 1.5 million tons/year.

The deposits are found in an area on the western side of the desert region lying between the coastal range and the Andes, generally in or just beneath bowl-shaped depressions in the foothills of the coastal range, at elevations ranging from about 4000 to 7500 feet, in a strip with a width of 5 to 10 miles, extending some 400 miles north and south.

The *caliche*, as the richer portion of the workable deposits is called locally, consists of cementitious material surrounding nodules of rock. The *caliche* occurs as a layer varying in thickness from 2 to 20 feet, and is generally covered with 1 to 12 feet of sandy overburden. When of lower grade, the material is known as *costra*. The *caliche* analyzes 7 to 12% $NaNO_3$, with a smaller percentage of KNO_3, 6 to 10% NaCl, 9 to 15% Na_2SO_4, with varying percentages of $CaSO_4$ and $MgSO_4$. It also contains small quantities of iodine, as $NaIO_3$, which is recovered and exported, and of boron, which must be removed if the nitrate is to be used as a fertilizer.

In earlier days, the ore was mined by hand, after breaking it up by blasting. It is said that the workmen were able to judge the quality of the *caliche* by crushing a little, throwing it on the glowing tinder, and noting its behavior. The nitrate content can, of course, be determined by standard methods of chemical analysis, such as the use of the nitrometer, in which the material suspended in sulfuric acid is reacted with mercury, which reduces the nitrate to nitric oxide, the volume of which is measured in a burette.

The origin of the nitrate deposits has been the subject of much theoretical speculation without reaching any unchellenged conclusion. The latest account, as given in the 1961 edition of Mellor,[3] suggests that the Chile *caliche* may have originated from botanical accumulations in a sea or large lake which slowly became isolated and ultimately dried up in the arid climate of the area. The process of daily evaporation and nightly cooling, combined with the different solubilities of sodium phosphates and sodium nitrate, may be sufficient to explain the relative absence of phosphate and, together with normal geological processes, the formation of *costra* or overburden.

The region where the nitrate beds occur is practically rainless,

[3] J. W. Mellor, *Comprehensive Treatise on Inorganic and Theoretical Chemistry*, Vol. 2, Suppl. 2, Wiley, New York and London, 1961.

and water for the operation has to be brought by pipeline across the desert from the Andes, 100 miles or so away. Fuel and all other supplies likewise have to be brought from great distances.

The principles of the method of extraction have scarcely changed since the early days of *caliche* mining. The ore is broken up with explosives, and the soluble salts are extracted with warm or hot water. Sodium nitrate is much more soluble at high temperatures than at low. Sodium chloride, on the other hand, increases little in solubility upon raising the temperature; in fact, in the presence of the nitrate, the solubility decreases. Consequently, NaCl does not deposit out to any great extent upon cooling and crystallization.

Details of the crude methods used in the early days have not been recorded. What is known as the Shanks process was introduced in 1876, and was used in the production of the greatest part of the nitrate exported from that time for some fifty years. For this process, the ore is broken up and crushed to about 2-inch lumps, and loaded into extraction tanks where it is boiled with mother liquor coming from the crystallizers. The strong liquor is clarified (with the addition of flour) and allowed to cool and deposit crystals of nitrate, which are drained, dried, and bagged for shipment. The residue from the first extraction is washed in batches with water advanced from successive stages, and finally discharged through doors in the bottom of the tank and sent to the dump for spent ore. The process had a high cost for fuel and for labor, and was economical, therefore, only on high-grade ore.

A more economical process was developed in the 1920s by the Guggenheim interests. It uses mechanical mining methods and mechanical treatment of the ore, which is crushed to a smaller size. The extraction is done at only about 40°C, and crystallization is carried to about 5°C, by using ammonia refrigeration. The solutions are filtered to remove slimes, and heat exchangers are used between the hot and cold solutions. Power is furnished by Diesel engines, and the exhaust and cooling water are used to supply the heat. Crystals are produced in suspension, and are drained in a mechanical thickener and centrifuge. Then they are melted and spray cooled for bagging and shipment.

It was mentioned earlier that the *caliche* contains varying amounts of potassium nitrate. During World War I, when supplies of potassium chloride ("muriate of potash") from Germany were cut off, the price went up from the level in 1914 of $35/ton to a peak of $500/ton. This gave a great incentive to the recovery of potash from the nitrate fields of Chile. Technologists at the plant which the Du Pont Company operated in Chile in that period worked out a method for producing a nitrate containing 25% of the potassium salt, by evaporating the mother liquor from the crystallizing pans, and the method, or a modification of it, is still in use. While small in terms of percentage, the production is considerable, running about 50,000 tons/year.

Before we leave the subject of saltpeter, we might take a closer look at the process for making potassium nitrate from sodium nitrate and potassium chloride. The process is based on the marked differences in the solubilities of the four salts involved, the two starting materials and the two produced by double decomposition ("metathesis"): namely, sodium chloride and potassium nitrate—though all four are classed as soluble salts. You may not have thought about it this way, since you may have associated double decomposition with such reactions as the precipitation of insoluble silver chloride from a solution of silver nitrate on the addition of potassium chloride.

It turns out that KNO_3 is more than ten times as soluble in water at 100°C as it is at 10°C, while NaCl increases in solubility only 10% over the same interval. You see that we have here the elements of a workable process. The raw materials, $NaNO_3$ and KCl, are dissolved in a quantity of boiling water which will take all the KNO_3 formed into solution, even though it leaves some NaCl in the solid state. On cooling to a low temperature, KNO_3 will be deposited, with only a little of the chlorides, and on repeated recrystallization from boiling water, it can be rendered substantially pure. You might like to try it for yourself. It makes a nice experiment, even for the home laboratory, involving no more danger than making a batch of candy. For this reason, I am including in the Appendix detailed, "cookbook," directions for carrying out the experiment, together with the essential solubility data, and a review of the theory on which it depends. There is one precaution to be observed: Filter papers soaked with nitrate solutions should not be left around, because they become flammable when they dry out.

This relatively simple process will also afford, if you care to try it, a typical example of the primary work that the industrial chemist or chemical engineer is called on to do: that is, to find out if the proposed process is likely to pay. Will the value of the product that can be sold leave enough over, after paying for the raw materials, for the wages of the plant operators, for the overhead related to wages and general running of the plant, for maintenance and repairs, and after provision for recovering the cost of the plant at the end of its useful life (through an amount set aside for depreciation): Will it leave enough over to offer a return on the money it will cost to build the plant? Do not forget that present Federal taxes on corporation profits will take approximately 50% of the profits. You will find data in the Appendix for making such a calculation, and this calculation will not give you any difficulty. You are not asked to estimate the cost of the plant: that is the work of the graduate chemical engineer. But it will take only elementary chemistry and arithmetic to find out how much you could afford to spend to build a plant to produce potassium nitrate. Incidentally, it will not take you long to figure out that if the yield of salable product is as low as in the experiment described, it will not pay to go into the business on any scale of operations.

Nitric Acid from Nitrates

After such an extended excursion into the traditional sources of nitric acid, let us turn now to the history of nitric acid itself. The account that follows is based largely on that of Mellor.[4]

The alchemists of the later Middle Ages were acquainted with the properties of nitric acid, which they called *aqua fortis* (strong

[4] J. W. Mellor, *Comprehensive Treatise on Inorganic and Theoretical Chemistry*, Vol. 8, Longmans, Green, London, 1928.

water) or sometimes *aqua valens* (powerful water), and they had several recipes for preparing it. The earliest known description of a method of making nitric acid appears in a twelfth-century work with the Latin title *De inventione veritatis* (The invention—or perhaps discovery—of truth), purportedly written earlier by the Arabian Geber. It says:

> Take a pound of Cyprus vitriol, a pound and a half of saltpetre, and a quarter of a pound of alum. Submit the whole to distillation, in order to withdraw a liquor which has a high solvent action. The dissolving power of the acid is greatly augmented if it be mixed with some sal ammoniac, for it will then dissolve gold, silver, and sulphur.[5]

Cyprus vitriol is presumably copper sulfate pentahydrate, and distillation of this mixture would give nitric acid. Addition of ammonium chloride would give what we still call by the alchemical name of *aqua regia* (royal water, from its ability to dissolve gold as well as silver).

It was the Dutch chemist Glauber (1603–1668) who first made nitric acid by distillation of saltpeter with sulfuric acid, or oil of vitriol, as he called it, and as it is sometimes called even today. We commemorate Glauber in the name we give to sodium sulfate decahydrate, "Glauber's salt." He deserves to be called a chemist, though the directions he gives in his book, published in 1658, for the preparation of nitric acid (see Figure 1.2) are couched in the metaphorical terms of the age of alchemy (the translation dates from 1689). It is hard to find how sulfuric acid gets into his formula. The "white swan" was a mixture of tin and saltpeter, in the vocabulary of the alchemists.

But having produced hydrochloric acid from common salt and sulfuric acid, he writes in a matter-of-fact manner in the same book: "Plainly after the very manner we have taught spirit of salt to be prepared, so may also *aqua fortis* be made. Instead of salt, take saltpetre, and you will have *aqua fortis*." And he speaks like a scientist rather than a practitioner of black arts when he observes: "I know well that ignorant laborators which

[5]Mellor, *op. cit.*, Vol. 8, p. 555.

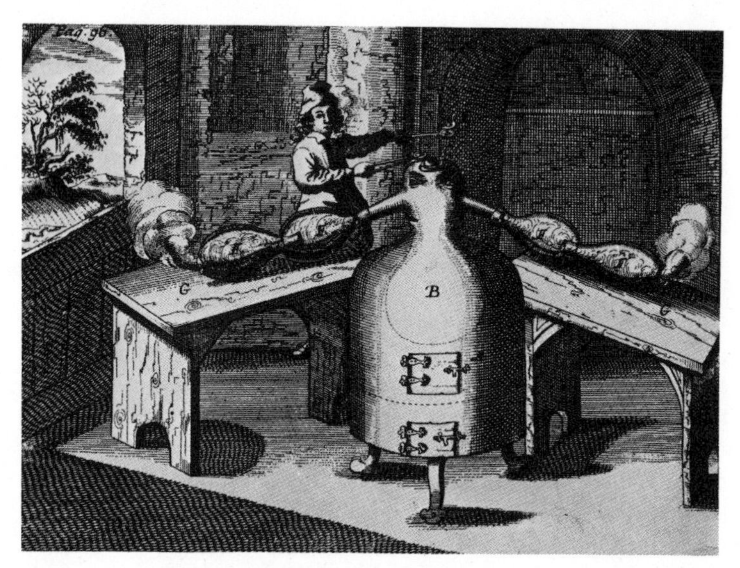

Figure 1.2 Iron man with two noses.

First a man is to be made of iron, having two noses on his head, and on his crown a mouth which may be opened and again close shut. This if it to be used for the concentration of metals is to be so inserted into another man made of iron or stone, that the inward head only may come forth of the outward man, but the rest of his body or belly may remain hidden in the belly of the exterior man. And to each nose of the head glass receivers are to be applied to receive the vapours ascending from the hot stomach. When you use this man you must render him bloody with fire to make him hungry and greedy of food. When he grows extremely hungry he is to be fed with a white swan. When that food shall be given to this iron man, an admirable water will ascend from his fiercy stomach into his head, and thence by his two noses flow into the appointed receivers; a water, I say, which will be a true and efficacious aqua-vitæ; for the iron man consumeth the whole swan by digesting it, and changeth it into a most excellent and profitable food for the king and queen, by which they are corroborated, augmented, and grow. But before the swan yieldeth up her spirit she singeth her swan-like song, which being ended, her breath expireth with a strong wind, and leaveth her roasted body for meat for the king, but her *anima* or spirit she consecrateth to the gods that thence may be made a salamander, a wholesome medicament for men and women.

From J. W. Mellor, *Comprehensive Treatise on Inorganic and Theoretical Chemistry*, Vol. 8, p. 556, Longmans, Green, London, 1928, with permission.

do all their work according to custom, without diving any further into the nature of things, will count me a heretic because I teach that the *aqua fortis* made of vitriol and saltpetre is of the same nature and condition as the *spiritus nitri* which is made without vitriol."[6]

In any event, the process which he employed for making nitric acid, by heating saltpeter with sulfuric acid, is the one that persisted in the laboratory and in industry right down to quite recent years, with only the substitution of sodium nitrate for potassium nitrate since the time it became available, not much more than 100 years ago. It is a process that you could carry out yourself, in your home or school laboratory. Many generations of chemistry students did carry out the experiment, and used to think it was fun, though it has disappeared from recent laboratory manuals, paralleling its obsolescence in industrial practice. You will find the simple directions for carrying it out in the Appendix.

According to an account dated 1771,[7] nitric acid was still being manufactured in glass retorts, 9-in. diameter and 12-in. height, holding 14 lb of niter. But practice was changing over to the use of an iron pot with an earthenware "pothead" connecting to earthenware receivers—not unlike Glauber's "iron man with two noses." The pot described in 1771 held 200 lb of niter. The essential features continued to be incorporated in equipment used for the production of nitric acid as long as it was made by this process. These essential features were dictated by the properties of the materials involved. What material of construction could you use to hold a charge of hot concentrated sulfuric acid? It was found that cast iron would stand up, at least for a useful number of charges. But for condensing nitric acid, nothing would withstand its action except ceramic or vitreous materials—dense earthenware or glass. (Gold, perhaps; but it was obviously too expensive, though it has been used in making "cp," chemically pure, acids for use as laboratory reagents.)

Limited as they were to these materials of construction for handling such highly corrosive chemicals, the industrial chemists of the eighteenth and nineteenth centuries had to exercise all their

[6] *Ibid.*

[7] Cited in A. Cottrell, *The Manufacture of Nitric Acid and Nitrates*, Gurney and Jackson, London, 1923, from which much of these paragraphs is taken.

ingenuity to put equipment together that would give any reasonably economic service. They were not called chemical engineers in those days, and maybe that is proper, as they were guided principally by rule of thumb, developed by long and often costly experience, a guideline not entirely superseded by the scientific background of the modern chemical engineer.

Without stopping to detail the successive steps in the evolution of this now obsolete process, let us pass to a description of the equipment as employed in the heyday of the art, about the time of World War I.

Chemistry of Nitric Acid Production

We ought, however, to review briefly the chemistry of the process. The production of nitric acid by distilling sodium nitrate with sulfuric acid is an essentially simple problem, and the reaction involved may be represented by the following equation:

$$\begin{array}{ccccccc} 2NaNO_3 & + & H_2SO_4 & = & Na_2SO_4 & + & 2HNO_3 \\ (2\times 85) & + & 98 & = & 142 & + & (2\times 63). \end{array} \qquad \text{(A)}$$

The volatile nitric acid is displaced from the system and may readily be condensed, while sodium sulfate remains in the reaction vessel.

In practice it is not advisable to work with the materials in the proportions just indicated because

1. The temperature required to complete the reaction is very high (approximately 900°C), and to work at such a high temperature would involve the following serious disadvantages:

a. Heavy wear and tear on the apparatus used.

b. High fuel consumption.

c. Loss of product due to decomposition of nitric acid to oxides and water and oxygen.

d. High nitrous acid content in the nitric acid produced, due to decomposition in the retort and absorption of the oxides in the acid in the condenser.

e. Production of weak acid as the result of the water produced in the decomposition.

2. The resulting sodium sulfate (mp 884°C) would readily set hard at the completion of the reaction and would be difficult to remove from its containing vessel.

By sacrificing a portion of the sulfuric acid represented in Equation A and working more nearly to the following equation:

$$\begin{aligned} NaNO_3 + H_2SO_4 &= NaHSO_4 + HNO_3 \\ 85 \quad + \quad 98 &= \quad 120 \quad + \quad 63, \end{aligned} \qquad (B)$$

the temperature for completion of the reaction need not exceed 200°C, and the residue remains sufficiently fluid to permit its being run freely from the reaction vessel. The reaction represented by Equation A involves the use of $NaNO_3$ and H_2SO_4 in the ratio 85:49, while the ratio for Equation B is 85:98. Prevailing practice was to use the proportions about 85 :80. The residue was then a mixture of sodium sulfate and sodium hydrogen sulfate; it was called "niter cake," merely from the process that gave rise to it, though it had no niter in it.

In the time that nitrate of soda was the source of nitric acid and when the demand for nitric acid was high, this niter cake, as an unavoidable by-product, had to be sold for whatever it would bring: it was a "drug on the market." It was used for its H_2SO_4 content in the manufacture of hydrochloric acid. The reaction of salt, NaCl, with sulfuric acid was generally carried all the way to Na_2SO_4, since no problem of decomposition is involved, and since the "muriatic acid" furnaces are capable of dealing with the dry solids, and since the reaction goes to completion below the melting point of Na_2SO_4. The product of that reaction was known as "salt cake." With the obsolescence of the old process for nitric acid, "niter cake" became in short supply, and it has now been found profitable to produce "synthetic niter cake," for the uses to which sodium hydrogen sulfate (bisulfate of soda) is put, by treating NaCl with sulfuric acid in much the same manner as nitrate of soda was treated, producing HCl as a by-product. The product is marketed as bisulfate of soda, containing the equivalent of 37 to 39% H_2SO_4.

This is just an illustration of the changes continually going on in the chemical manufacturing industry.

The chemistry of the process, as outlined here, and the physical properties of the raw materials and products determine the essential features of the equipment necessary and the method of carrying out the reaction.

Commercial Production of Nitric Acid from Nitrate of Soda

In its final state of development, the nitric acid plant included: a battery of retorts connected in pairs to condensers; absorption towers for the recovery of uncondensed vapors; a variety of

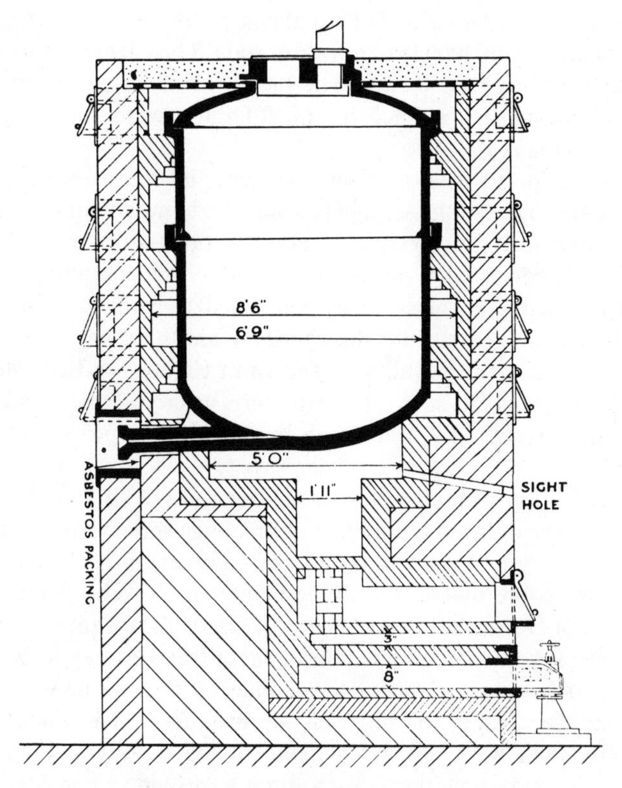

Figure 1.3 Retort setting, Gretna.

From Allin Cottrell, *The Manufacture of Nitric Acid and Nitrates*, p. 57, Gurney & Jackson, London, 1923, and D. Van Nostrand Company, New York, 1923, with permission.

accessory apparatus such as charging devices for nitrate and acid, storage systems for the weak and strong portions of the acid produced, and mixing tanks if mixed acid is to be made (mixtures of nitric and sulfuric acids, as used in nitration processes).

The retort was a large cast iron vessel, cast in two or three sections, in a brickwork setting, arranged for coal, or alternatively for oil or gas firing. Gas firing is shown in the illustration of the retort employed at the large wartime British plant at Gretna (see Figure 1.3). Note the tap hole for the discharge of the niter cake. During a run, this is plugged with clay, and a solid ("blank") flange is fastened over it. The usual charge for a retort of the size shown was about 4000 lb of nitrate of soda. The life of a retort was only about 500 to 600 runs. The metal thickness was generally 2 in. Corrosion takes place, but most failures were said to result from cracking.

Connection to the condenser was made by means of earthenware pipes. The condenser most used in the United States was of a type developed by Dr. W. A. Hart, a professor at Lafayette College. It consisted of one or more stands, each containing fourteen inclined 3-in. glass tubes connected to baffled stoneware headers, arranged so that the vapors made three passes. Water was trickled over the tubes, as shown in Figure 1.4. Each stand provided about 50 sq ft of condensing surface. Tube breakage was high, reportedly about one hundred tubes per retort per year.[8]

Substitution of high-silicon iron, Duriron (13 to 16% Si), for earthenware reduced the danger of breakage somewhat, but this, too, is a pretty brittle material of construction and not very resistant to thermal shock. Some plants employed S-bend condensers of Duriron, others S-bends of fused silica ware.

There is some decomposition of the nitric acid vapors at the temperatures reached during the distillation, and it is necessary to recover the oxides of nitrogen so produced. The chemistry involved will be taken up in detail in the following chapters. Suffice it to say now that it is required to provide a considerable

[8]F. C. Zeisberg, *Chemical and Metallurgical Engineering*, **24**, 443–445 (1921).

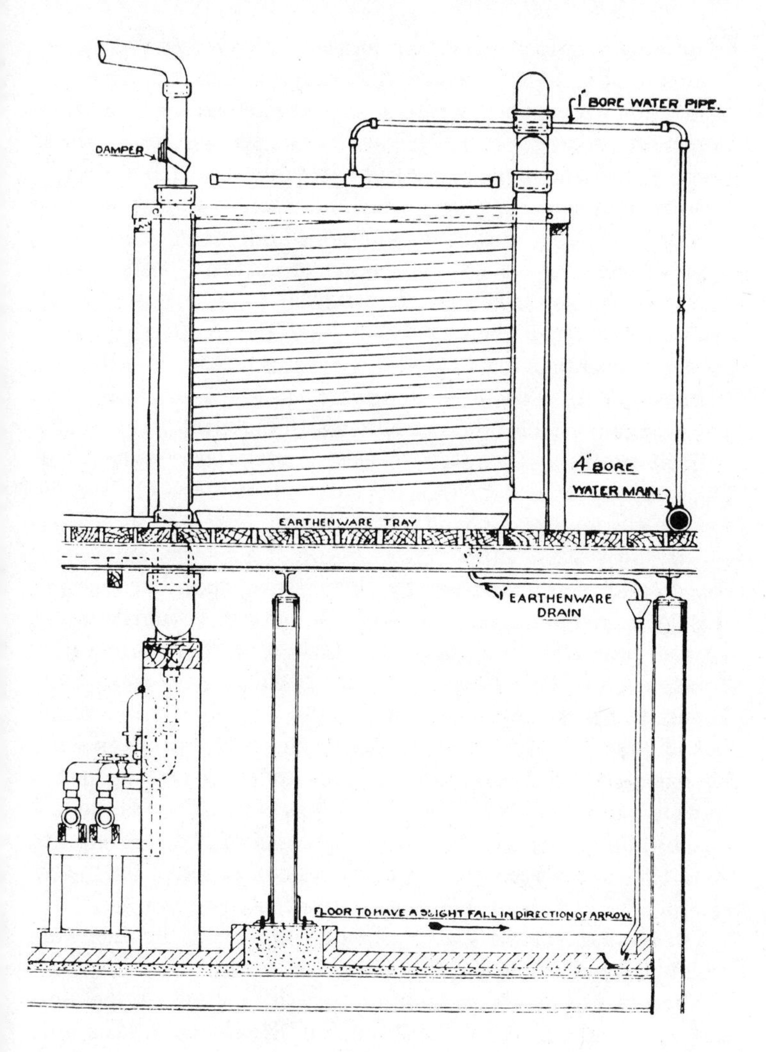

Figure 1.4 Hart's condenser.

From Cottrell, *op. cit.*, p. 67, with permission.

volume in the absorption system, and successive stages of counter-current treatment with water. According to Zeisberg,[9] the first absorption towers using water were built at Hercules, California, from sewer pipe set vertically, with broken beer bottles for tower packing. The acid that drained from the bottom of each tower was allowed to flow into a glass carboy and when this was full, it was carried upstairs by workmen and dumped into the top of the next tower. Later, the towers were made of chemical stoneware, a dense, impervious earthenware. A typical tower consisted of six sections 30 in. in diameter by 30 in. high. Lump quartz was first used for tower packing, to expose a large surface to the gases, but formed packing of chemical stoneware was found better. Finding a suitable material for the saucer in which the tower rests posed a difficult problem. Granite was used, and "volvic stone"; but Duriron was used here also. Circulation of the acid over the towers was another difficulty. Egg-shaped vessels of stoneware, called "blowcases," in which a charge of acid was allowed to collect, and then discharged by air pressure, upon opening and closing the proper valves, were used. These vessels could be rigged up for automatic operation, in which case they were called "pulsometers." But these were not very safe, on account of the fragile materials of construction.

The "air lift" was therefore used instead. The airlift operates by balancing, hydrostatically, an air-liquid mixture against a shorter column of liquid; successive slugs of liquid are raised in a pipe by admitting air in the form of bubbles at the base. It requires 50 to 60% submergence of the base, and consequently the towers supplied by this means were set on an elevated platform.

Finally, flat iron pans were provided into which the niter cake could be run and allowed to solidify.

The usual cycle of operation of a retort as shown in the illustration was about 14 hr. The strength of the distillate was a little low at first, due to water in the nitrate and the sulfuric acid (92.5% was thought to be the optimum concentration of H_2SO_4) but soon rose to about 97%, after which it fell off gradually. The rate

[9] *Ibid.*

of distillation reached a maximum about 3 hr after heat was applied, and then decreased. Soon after firing, the charge was said to foam, and the foam depth increased until just before the end of the reaction, when the retort became practically full of foam. The foam level, after reaching this point, dropped abruptly, serving as an indication that the reaction was complete. Another indication of complete reaction was the appearance of purple iodine vapors in the absorbers, from the small amount of iodate in the Chilean nitrate.

The acid was run from the condensers, usually by gravity, in glass pipe lines supported in wooden troughs, and then loaded for shipment in glass carboys—once a familiar sight around a chemical plant. If the acid was intended for nitration, it was run directly into a charge of sulfuric acid, contained in a steel mixing tank provided with means for circulation and cooling.

The nitric acid plant, with the fires for heating the retort, and the pans of niter cake cooling, and stoneware and glass lines carrying strong nitric acid and nitrogen oxides, always in danger of breakage, was not a pleasant place in which to work, as you can perhaps imagine. Consequently there were few regrets when it was rendered uneconomic, and consequently obsolete, beginning about 1915 (in Germany) and about 1925 (in the United States), by the introduction of synthetic ammonia and the ammonia oxidation process. The last plant was dismantled by about 1940.

[J. R. Partington, in *A History of Greek Fire and Gunpowder* (New York, Barnes & Noble, 1960), casts doubt upon the accuracy and even the authenticity of Bacon's anagram. Note added in proof.]

CHAPTER TWO

THE CHEMISTRY OF NITROGEN OXIDES

The present-day process for making nitric acid starts with ammonia, not sodium or potassium nitrate. A review of the chemistry of the process will give you an understanding of the limitations within which the operations must be carried out, and of the requirements to which the equipment used must conform.

The ammonia oxidation process, in outline, is relatively simple. It starts with a single pure compound, plus air and water, and ends up with a nearly quantitative yield of another pure compound (in water solution), with no by-products of any consequence. It has, just the same, some points of unique interest which may not have been brought out in your studies thus far.

A single equation can be written for the over-all process

$$NH_3 + 2O_2 = HNO_3 + H_2O,$$

but actually the process involves three stages. The first is a gas-phase catalytic reaction, the primary oxidation of ammonia to nitric oxide,

$$4NH_3 + 5O_2 = 4NO + 6H_2O,$$

carried out with the aid of a platinum or platinum-alloy catalyst, at a temperature of around 900°C, with a very short contact time.

The second is a homogeneous, noncatalytic, gas-phase reaction between nitric oxide and additional oxygen to produce nitrogen dioxide,

$$2NO + O_2 = 2NO_2.$$

This is a relatively slow reaction, and is actually the limiting, or rate-determining, step in the sequence of reactions. Though we need not dwell on the point right now, nitrogen dioxide exists, at ordinary temperatures, in equilibrium with a dimeric form, dinitrogen tetroxide,

$$2NO_2 = N_2O_4.$$

It will be convenient to refer to this equilibrium mixture as "nitrogen peroxide."

Finally, this nitrogen peroxide is absorbed, or more exactly speaking, treated in countercurrent fashion, with water. Each successive stage produces one molecule of nitric oxide for every two molecules of nitric acid,

$$3NO_2 + H_2O = 2HNO_3 + NO.$$

This NO must be oxidized back to the peroxide state for further reaction.

Because several oxides of nitrogen are involved in the successive steps in the process, let us take a closer look at the chemistry and the physical properties of the oxides of nitrogen.

It is worth noting, first of all, that the chemical inertness of "free," or elementary, nitrogen does not imply that nitrogen is at all an inactive element. As Michael Faraday once said: "What of nitrogen? Is not its apparent simplicity of action all a sham?" The inertness of nitrogen results from the high energy of dissociation of molecular nitrogen into nitrogen atoms, amounting to 226,000 cal/mole N_2 (at 25°C).

The element nitrogen, with five electrons in the outer shell of the atom, two *s* and three *p* electrons, would be expected to enter into combination in valence states -3, $+3$, and $+5$. For example, nitrogen has a valence of -3 in ammonia, NH_3. The other valence states are exhibited in oxides; for example, in N_2O_3, the anhydride of nitrous acid, HNO_2, nitrogen has a valence of $+3$; in N_2O_5, the anhydride of nitric acid, HNO_3, the valence is $+5$. But there are other oxides, some of key importance to the understanding of the ammonia oxidation process.

Nitrous Oxide

First there is nitrous oxide, N_2O, made as long ago as 1793 by gentle heating of ammonium nitrate, and made commercially to the present day by the same reaction. You could make some yourself, if you are careful about it, by cautiously heating some (chemically pure, cp) ammonium nitrate in a Pyrex test tube with a glass tube connection leading to a gas collecting bottle set in a trough of water. Caution: Use eye protection, and disconnect the tubing before the test tube starts to cool. The reaction, probably complex in its mechanism, is stated simply enough:

$$NH_4NO_3 = N_2O + 2H_2O$$

Nitrous oxide has been known, since 1800, when Sir Humphry Davy reported his extensive experiments, as "laughing gas." It is of such a low degree of toxicity that it has long found use as an anesthetic, in dentistry and in minor operations.

But nitrous oxide is not encountered in the manufacture of nitric acid by ammonia oxidation, at least not if the reaction is properly managed. It is formed if a mixture of ammonia and air is passed over platinum at a low red heat, about 500°C, at a slow rate. This method, however, would result in a loss of yield in a process directed toward the production of nitric acid, but such loss can be minimized and substantially eliminated by working at higher temperatures, on the order of 900°C, at shorter contact times.

Nitrous oxide is a colorless gas, with a boiling point of −88.48°C at atmospheric pressure, and a freezing point only 2°C lower, −90.82°C. The critical temperature, the temperature above which it cannot be liquefied, is 36.5°C, and the critical pressure, the pressure required to cause it to liquefy at that temperature, is 71.7 atmospheres (1054 lb/sq in. abs).

Nitrous oxide supports combustion, in certain cases almost as well as oxygen. It can be considered hypothetically as the anhydride of hyponitrous acid, but that is a rare compound, which does not even exist in the free state. In any event, it has no relation to nitric acid, and we shall not consider it further.

Nitric Oxide

Nitric oxide, on the other hand, demands our attention. It is the primary product in the process of oxidizing ammonia with the objective of making nitric acid: under proper conditions it is practically the sole product, with, of course, the water produced in the reaction,

$$4NH_3 + 5O_2 = 4NO + 6H_2O.$$

Nitric oxide was discovered by Priestley in 1774, who produced it by the action of nitric acid on copper. Its composition was established by Sir Humphry Davy, about 1800.

It is conveniently produced in the laboratory by adding a solution of ferrous sulfate, acidified with sulfuric acid, to a solution of sodium nitrate:

$$2FeSO_4 + 2NaNO_2 + 2H_2SO_4 = Fe_2(SO_4)_3 + 2NO + 2H_2O + Na_2SO_4.$$

Nitric oxide has a boiling point of -151.74°C and a melting point of -163.51°C. In the pure state, nitric oxide is a colorless gas; the critical temperature is -94°C, and the critical pressure is 65 atm. The gas is completely monomeric, with a density corresponding to the formula NO. There is evidence that in the liquid and solid states there is some degree of dimerization to N_2O_2. It does not react with water, and is only slightly soluble in it.

Nitric oxide is reactive, however. It reacts at ordinary temperatures with oxygen, and we shall need to consider this reaction more closely. It also reacts with chlorine and with bromine, forming nitrosyl chloride and nitrosyl bromide, respectively; and with sulfuric acid, forming a crystalline compound, nitrosyl sulfuric acid, or nitrosyl hydrogensulfate, an intermediate in the old lead chamber process for making sulfuric acid.

It is the reaction of nitric oxide with oxygen that is of concern to us, since it is the slowest, and therefore the rate-determining, step in the sequence of reactions leading to the production of nitric acid. It will be more illuminating to discuss the mechanism of this reaction—the details of the succession of molecular changes through which the reactants pass—after we have learned something about the higher oxides of nitrogen. For the moment, let us consider only the significance of the over-all rate factors.

The reaction of nitric oxide with oxygen is unique in several respects. It goes spontaneously at room temperature, without requiring any catalyst or initiating spark. But contrary to most spontaneous, homogeneous, gas-phase reactions, it goes rather slowly, at an easily measurable rate. And contrary to most spontaneous reactions, it goes more slowly as the temperature is raised, whereas you may have been told that the rate of many reactions doubles for every 10°C rise in temperature.

Moreover, it goes – effectively – as written, as a termolecular (three-molecule) reaction:

$$2NO + O_2 = 2NO_2.$$

The rate at which it goes actually is proportional to the square of the concentration of the NO multiplied by the concentration of the oxygen. If you have learned anything about the kinetic theory of gases in your physics or chemistry courses, you will recall that the number of collisions between pairs of molecules can be estimated, but that three-body collisions are so rare that they can be practically neglected. Nevertheless, measurements of the rate of this reaction R (under the conditions of interest to us) do conform to the mathematical expression

$$R = \frac{dp_{NO_2}}{dt} = kp_{NO}^2 p_{O_2}$$

Whether or not you have studied calculus, do not be frightened by this differential equation. The solution ("integration") will presently be performed for you. Let us first introduce a few symbols to make the equations look a little simpler.

Let a = initial partial pressure of NO, atm
b = initial partial pressure of O_2, atm
c = partial pressure of NO_2 at time t, atm
t = time from start of reaction, sec
x = c/a, fraction oxidized at time t.
k = reaction rate constant.

$$\frac{dc}{dt} = k(a-c)^2\left(b-\frac{c}{2}\right),$$

$$a\frac{dx}{dt} = k(a-ax)^2\left(b-\frac{ax}{2}\right).$$

Assuming that the reaction proceeds without any appreciable back-mixing of products with the entering gases, we get, after some simplification, the following expression for the time t required to reach a given degree of oxidation:

$$t = \frac{2}{k(2b-a)}\left[\frac{1}{a(1-x)}-\frac{1}{a}-\frac{1}{2b-a}\log_e\frac{2b-ax}{2b(1-x)}\right].$$

This still looks rather formidable, but for many cases the term containing the natural logarithm can be neglected, and the resulting expression, still further simplified, is one you would certainly be prepared to deal with:

$$t = \frac{2}{ak(2b-a)}\left(\frac{x}{1-x}\right).$$

Better than anything else, working out a couple of examples with the aid of these formulas will give you an idea as to how physicochemical data can be applied to industrial problems. At the same time you can gain an insight into the essential controlling factors in the process of making nitric acid by the oxidation of ammonia.

A number of investigators have made measurements of the rate constant of this reaction, generally checking the classic results of Bodenstein.[1] I made a couple of high-spot tests myself in 1925 (unpublished), and arrived at the same answer. It is amazing how much controversy arose before that time as to the rate factors, but there is no use going into that now.

An expression that represents Bodenstein's results is

$$\log k = (641/T)-0.725,$$

[1] M. Bodenstein and P. Lindner, *Zeitschrift für physikalische Chemie*, **100**, 105 (1922).

where T is the temperature ($°K = °C+273.1$). For a temperature that might occur in an absorber in the ammonia oxidation process, 43°C, 316°K, $k = 20.0$.

Now let us calculate the time necessary to secure 95% oxidation to NO_2 under two sets of conditions. First let us take the conditions as atmospheric pressure, with 10% NO by volume at the outset and 7% O_2, $a = 0.10$, $b = 0.07$, as might be found in the gases coming from an ammonia oxidizer.

Substituting these values in the (complete) expression for t, we obtain

$$t = \frac{2}{(20)(0.14-0.10)}\left[\frac{1}{(0.10)(1-0.95)} - \frac{1}{0.10} - \frac{1}{(0.14-0.10)}\log_e\frac{(0.14-0.095)}{(0.14)(1-0.095)}\right],$$

and $t = 359$ sec, or approximately 6 min. (Working through this example, you would note in this case that the logarithm term could not be neglected.)

Now let us see how long it would take for the same extent of oxidation under a total pressure of 8 atm, all other conditions the same. Now $a = 0.80$, $b = 0.56$, and substituting and solving,

$$t = \frac{2}{20(1.12-0.80)}\left[\frac{1}{(0.80)(1-0.95)} - \frac{1}{0.80} - \frac{1}{(1.12-0.80)}\log_e\frac{(1.12-0.76)}{(1.12)(1-0.95)}\right],$$

and $t = 5.6$ sec.

It was set up that way, but you see it comes out right; the time required at 8 atm is just $\frac{1}{64}$th of that at 1 atm, conforming to the relationship for a third-order reaction, the rate being proportional to the square of the pressure, as you would read in any treatment of chemical reaction kinetics.

You see then, of what importance the rate of this one reaction is in the process of making nitric acid by ammonia oxidation. Another factor that must always be taken into account in any reaction carried out on an industrial scale is the heat effect, the *thermochemistry* of the reaction. We assumed a temperature in working out the example, but this assumption would require the chemical engineer to provide sufficient heat transfer surface to maintain a temperature of the gas in the reaction space at 43°C, with water available at, say 30°C (86°F), or air at about the same temperature. The chemical engineer has to calculate first the amount of heat that would be given out in a reaction, and how high this heat would raise the temperature of the gas mixture (or liquid, as the case might be).

In tables of thermochemical data you would find the heat of formation of nitric oxide given as +21,600 cal/g mole, meaning that 21,600 calories would have to be supplied to produce 1 g mole of NO from the elements. The heat of formation of NO_2 is given as +8060 cal/g mole. We obtain the heat of reaction by subtracting the heat of formation of NO from that of NO_2 (that of the oxygen being taken again as zero). We get −13,540 cal/g mole, which we interpret as the amount of heat given off when 1 g mole of NO_2 is formed from NO and O_2. Now let us calculate the amount of heat that would be given off in letting 100 g moles of gas containing 10% by volume of NO and 7% O_2 react to 95% completion of the reaction. We shall assume the remainder of the mixture to be nitrogen. To a very good approximation, sufficient for a preliminary engineering estimate, we may take the molar heat capacity of the gas mixture to be 7.0 cal/(g mole) (°C). In the gas mixture there will be 10 g moles of NO, and we will make the calculation for the production of 9.5 g moles of NO_2. The amount of heat released is given at once as $9.5 \times 13{,}540 = 129{,}000$ cal. Then we can readily calculate the temperature rise of the gas mixture if no provision were made to remove this heat. It would be simply $129{,}000/(7)(100) = 184$°C. Since you have just been reading that the reaction slows down as the temperature is raised, you will see that something has to be done about this potential temperature rise.

In fact, it will become more and more evident that the important things to take care of in the industrial application of the ammonia oxidation process for making nitric acid are the two things we have just looked at: provision for time for the reaction between

NO and O_2, and removal of the heat of reaction at all stages.

Another aspect of the reaction between nitric oxide and oxygen that we ought to take a look at is the equilibrium, though it turns out not to be of as much importance to us as the two we have just considered.

We assumed that the reaction as written would eventually go to completion in the forward direction

$$2NO + O_2 = 2NO_2$$

if we gave it enough time; and at ordinary temperatures this is practically true. At moderately elevated temperatures, however, the reverse reaction, as in most exothermic reactions, becomes dominant, and nitrogen dioxide dissociates, and nitric oxide is not oxidized. Expressed in the form corresponding to the Law of Mass Action,

$$K_p = \frac{p_{NO_2}^2}{p_{NO}^2 p_{O_2}}.$$

Values of this equilibrium constant were also determined by Bodenstein.[2]

Representative values are given:

t, °C	K_p
38	1×10^{11}
200	1×10^{5}
400	1×10^{1}
600	1×10^{-1}
800	1.7×10^{-3}
1000	1.2×10^{-3}

It can be concluded from this list that at the temperature of the ammonia oxidation reaction, 900°C more or less, nitric oxide is the oxide formed, since K_p there is much less than unity. On cooling the gases to temperatures where they could be brought into contact with liquid water, 200°C or below, NO_2 is stable, the value of K_p being 10^5 or more, and the dissociation into NO and

[2]M. Bodenstein, *Zeitschrift für Elektrochemie*, **34**, 183 (1918).

O_2 does not have to be taken into account as a practical matter.

We should not leave the discussion of the chemistry of nitric oxide without referring to still another feature, namely, its stability with regard to dissociation into the elements. While not important to the ammonia oxidation process for making nitric acid, such stability is vital to the electric arc process for nitrogen fixation and to the later thermal process (to be discussed in Chapter Five). Thermodynamically, as an endothermic compound, it is unstable at ordinary temperatures—as, indeed, are N_2O and NO_2. But its rate of dissociation into the elements is so slow as to be entirely negligible. Writing the equation for its formation as

$$N_2+O_2 = 2NO,$$

and correspondingly writing the equilibrium constant for the formation of NO

$$K_p = \frac{p_{NO}^2}{p_{N_2}p_{O_2}},$$

we find the following values tabulated,[3] calculated by thermodynamics, except the one at 2402°C, which is based on experimental data:

t, °C	K_p
25	5.3×10^{-31}
727	8.0×10^{-9}
1538	1.3×10^{-4}
2402	3.4×10^{-3}
3227	4.4×10^{-2}
4727	2.8×10^{-1}

As far as it concerns us here, however, the rate of the reaction even at the highest temperature reached in the ammonia oxidation reaction is so low as to be negligible.

We find the following values reported[4] for the time required to

[3] H. H. Sisler in M. C. Sneed, R. C. Brasted, and J. L. Maynard, Eds., *Comprehensive Inorganic Chemistry*, Vol. 5, Part I, Van Nostrand, Princeton, N.J., 1956.

[4] E. B. R. Prideaux and H. Lambourne in J. A. N. Friend, Ed., *Textbook of Inorganic Chemistry*, Vol. 6, Part 1, Griffin, London, 1928.

decompose halfway to the equilibrium concentration (in the absence of a catalyst):

627°C	1227°C	1627°C
7350 sec	200 sec	1 sec

These values can be compared with the time required to form half of the equilibrium concentration from the elements

1227°C	1627°C
108,600 sec	125 sec

Since the maximum temperature reached in the ammonia oxidation process is only about 900°C, and that only for a fraction of a second, we can safely conclude that decomposition of the nitric oxide formed in the reaction, even though it is accelerated by platinum,[5] is not of real concern to us here.

Before we conclude the discussion of nitric oxide, we might take note of the fact that its other gas-phase reactions, with chlorine and bromine, at ordinary temperatures, and with hydrogen at around 800°C, are also relatively slow, and also follow the kinetic equation for a termolecular process.

Nitric oxide is toxic, if only for the reason that by the time it reaches the lungs it would be converted to the dioxide, which is quite toxic. In very low concentrations it is merely irritant. It is believed to contribute to the irritant effects suffered in smog, prevalent in such localities as Los Angeles. Its presence may be due either to its formation in the upper atmosphere, by photochemical action of sunlight on nitrogen and oxygen in the air, or to nitrogen fixation at the temperatures reached in automobile cylinders to produce nitric oxide that is emitted in the exhaust.

Dinitrogen Trioxide, N_2O_3

Let us now leave the discussion of nitric oxide and take up the higher oxides in succession.

The first of these is dinitrogen trioxide, or nitrogen (III) oxide, in the nomenclature of systematic inorganic chemistry, more

[5] L. Andrussow, *Zeitschrift für angewandte Chemie*, **39**, 321–332 (1926).

conveniently referred to by its formula, N_2O_3. While recognized as the anhydride of nitrous acid, the very existence of this compound as a chemical individual has been the subject of controversy. Indeed, it can hardly be said to have an independent existence in the vapor phase because it dissociates extensively into nitric oxide and "nitrogen peroxide" (as we are calling the equilibrium mixture of NO_2 and N_2O_4). It will be necessary, therefore, to anticipate some of the discussion of this still higher oxide in order to understand the chemistry of N_2O_3.

Knowing as much as has now been clarified by modern physicochemical methods, we can easily see how the uncertainty as to the existence of N_2O_3 persisted for so long.

It is not hard to produce a mixture of gases with the empirical composition corresponding to N_2O_3 by mixing equal volumes of nitric oxide and nitrogen peroxide. At room temperature, the mixture will not show any decrease in volume and will still look, to the eye, as brown as it would if the nitrogen peroxide had been diluted with nitrogen instead of nitric oxide. You would think you were justified in deducing that there had been no reaction between the gases. But on refrigerating the mixture, you would not get the yellow liquid corresponding to nitrogen peroxide but a liquid of deep blue color (perhaps tinged slightly green at this actual concentration), which would freeze eventually to a light blue solid, and you would conclude that in the liquid and solid states you had a new compound, presumably N_2O_3.

Now, if we anticipate the data on the equilibria involved in the decomposition of N_2O_3 and of N_2O_4, it could be shown that the actual compositions before and after mixing equal volumes of nitric oxide and nitrogen peroxide, at 27°C, at atmospheric pressure, would be as follows for 100 volumes of each component:

	N_2O_4	NO_2	NO	N_2O_3	Total
Before	68	32	100	—	200
After	62	38	94	6	200

It just happens that under these conditions the dissociation of the N_2O_4 upon dilution exactly makes up for the association of NO_2 and NO to form N_2O_3.

The properties of N_2O_3 in the gas phase are characterized principally by the equilibrium relationship for its dissociation into NO and NO_2. The strong absorption of NO_2 in the visible region of the spectrum makes it fairly easy to investigate by photometric methods the equilibrium relationships existing between it, on the one hand, and N_2O_4 and N_2O_3 on the other, both of which are colorless, that is, with no appreciable absorption in the visible region, though they absorb in the ultraviolet. (See Figure 2.1.)

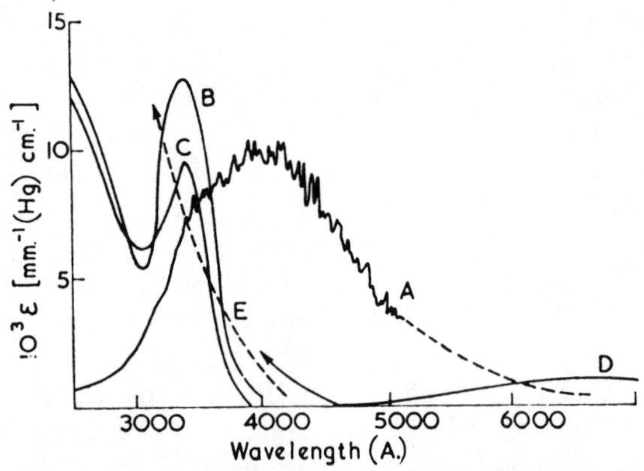

Figure 2.1 Spectrum of certain oxides of nitrogen.

A, nitrogen dioxide. B, dinitrogen tetroxide in cyclohexane. C, dinitrogen tetroxide in the gas phase. D, dinitrogen trioxide in toluene at —80°. E, dinitrogen trioxide in the gas phase (the curve from the lower arrow head increases rapidly to > 130).

From I. R. Beattie, *Progress in Inorganic Chemistry*, p. 22, F. A. Cotton, Ed., Interscience, New York, 1963, with permission.

The equilibrium constant for the reaction $NO + NO_2 = N_2O_3$, written as

$$K_p = \frac{p_{N_2O_3}}{p_{NO}p_{NO_2}},$$

has the following values, in atmospheres, as a function of temperature[6] (there is a small correction for total concentration which we will disregard in this discussion).

[6]J. C. Treacy and F. Daniels, *Journal of the American Chemical Society*, **77**, 2033 (1955).

t, °C	K_p
25	0.475
35	0.272
45	0.145

Without detailing the calculation, we can estimate, for example, what the equilibrium concentration of N_2O_3 would be in a gas containing 3.9% NO and 3.4% NO_2 by volume (46.5% oxidized, we would say), at a temperature of 35°C, and a total pressure of 1 atm. The concentration works out to 0.035% by volume, or about 0.5% of the total oxides present. You can see then that we are safe in assuming that for practical purposes we can neglect the presence of N_2O_3 under the conditions of the ammonia oxidation process.

At higher total pressures, it is true that the fraction existing as N_2O_3 would be higher, almost in direct proportion to the pressure. But even at 8 atm, the fraction combined as N_2O_3 would only be 4%, and we can still neglect it.

In the gaseous state N_2O_3 is colorless, but a gas mixture of this equivalent composition condenses to a deep blue liquid somewhere below −40°C. It freezes at −100.7°C, but forms with liquid N_2O_4 a eutectic (a mixture of constant composition on melting) of empirical composition $NO_{1.56}$ melting at −106.2°C. The vapor above a liquid of composition N_2O_3 consists of almost pure nitric oxide. A review by Beattie[7] should be consulted for further information about N_2O_3.

Nitrogen Peroxide, NO_2 or N_2O_4

We come next to nitrogen (IV) oxide, nitrogen dioxide, or in its dimeric form, dinitrogen tetroxide, and in the equilibrium mixture of the two forms, called, for convenience, nitrogen peroxide (though, having no oxygen-oxygen linkage, it is not a true peroxide). We have already touched on some of its properties, but it is worth repeating that this oxide is of major importance in

[7] I. R. Beattie in *Progress in Inorganic Chemistry*, F. A. Cotton, Ed., Vol. 5, pp. 1–26, Interscience, New York, 1963.

making nitric acid by ammonia oxidation. It is the form that finally reacts with water to produce nitric acid.

In 1670, William Clark described the properties of the gas we know as nitrogen dioxide, calling it, from its reddish color, "the flying dragon." Priestley prepared it in purer form in 1777, and somewhat later Dulong and (independently) Gay-Lussac established its composition.

Nitrogen dioxide can be prepared in the laboratory by the action of concentrated nitric acid on copper:

$$Cu + 4HNO_3 = Cu(NO_3)_2 + 2NO_2 + 2H_2O,$$

or by heating lead nitrate:

$$2Pb(NO_3)_2 = 2PbO + 4NO_2 + O_2.$$

It also results from the oxidation of organic compounds with nitric acid, and is found in the products of decomposition of organic nitrocompounds. As we are concerned with it, of course, it is produced by the oxidation of nitric oxide. We shall describe that process a little later.

The heat of formation of the dioxide NO_2, from the elements, as given previously, is +8060 cal/g mole (reactants and products at 25°C), making it still an endothermic compound. The heat of formation of the tetroxide N_2O_4, is +2420 cal/g mole, only slightly endothermic. Subtracting the heat of formation of two moles of NO_2 from that of one mole of N_2O_4, the reaction

$$2NO_2 = N_2O_4$$

is then exothermic, to the extent of 3220 cal/g mole, and this is another increment of heat for which the chemical engineer must be careful to account in the industrial development of the ammonia oxidation process.

The equilibrium between these two forms is the dominant feature of the properties of this oxide. Equilibrium is very rapidly attained in either direction, and for all practical purposes the reaction may be considered as being in the equilibrium state. The degree of association is high but varies markedly over the temperature range of interest to us. Writing the equilibrium constant as $K_3 = p_{N_2O_4}/(p_{NO_2})^2$, Bodenstein[8] obtained the following values:

t, °C	K_3, atm^{-1}
0	65
20	12
40	2.7
60	0.67
80	0.20

To see what this means under conditions we might encounter in the ammonia oxidation process, we can readily set up an algebraic equation to solve for x, the fraction of the nitrogen peroxide in the monomeric form at equilibrium at a given temperature. This equation is

$$x = \frac{p_{NO_2}}{p_{NO_2} + 2p_{N_2O_4}}$$

for a given total pressure P and value of $(p_{NO_2} + 2p_{N_2O_4})/P$. For values of this equivalent mole fraction of peroxide equal to 0.10, as might be found in ammonia oxidizer gases after separation of the water produced, and completion of the oxidation of the nitric oxide, we can calculate values of x at different temperatures and pressures.

TABLE 2.1
Fraction of Total Peroxide as NO_2, x
Mole Fraction of Peroxide in Gas, 0.10

t °C	x	
	1 atm	8 atm
0	0.250	0.092
24	0.533	0.245
80	0.950	0.81

[8]M. Bodenstein, *Zeitschrift für Elektrochemie*, **34**, 183 (1918).

If, as will appear shortly, it is the dimeric form N_2O_4, not the monomeric NO_2, for which the fraction is given in the Table 2.1, that reacts with water to form nitric acid, you can see how important it would be in practice to strive for low temperatures and high pressures in nitric acid manufacture. We see here another example of the industrial application of experimental measurements made in the physical chemistry laboratory.

The rapid change in degree of association of nitrogen (IV) oxide can be readily observed visually without elaborate instruments. Nitrogen dioxide, NO_2, is strongly colored, while the tetroxide N_2O_4, is almost colorless. The absorption spectra are shown in Figure 2:1. Remembering that the visible spectrum runs from about 3800 to 8000 Angstrom units (abbreviated as Å), violet to red, and remembering the laws of optics, that absorption in the blue region would leave red and yellow coming through, you can see why the gas looks red-brown. A bulb filled with nitrogen peroxide gas looks light brown at low temperatures, say 10°C, but on heating to 150°C, it looks almost black.

The peroxide is easily liquefied. The boiling point at atmospheric pressure is 21.15°C, and the liquid freezes to a light-colored solid at −11.2°C. The critical temperature is 158.2°C, and the critical pressure is very close to 100 atm.

Nitrogen dioxide is toxic. It affects the lungs and respiratory tract, though the symptoms may be delayed. The "maximal acceptable concentration" is given as 5 ppm (parts per million, by volume) for continuous exposure during a working day (8 hr), and under 25 ppm for 3 to 5 minutes.[9] The odor threshold is about 5 ppm, so if you can smell it, you had better find fresh air. If the brown fumes are visible, keep away from them: the visual threshold is 75 to 100 ppm.

The most characteristic reaction of nitrogen peroxide, and the one of most importance industrially, is its reaction with water, but we will delay consideration of that until we have discussed the higher oxides.

[9] American Standards Association, Z. 37.13 (1962).

Higher Oxides

These higher oxides are not of industrial importance, but a knowledge of certain of their properties will aid our understanding of the reactions that are important to us.

Dinitrogen pentoxide, N_2O_5, the next higher oxide, is the anhydride of nitric acid

$$N_2O_5 + H_2O = 2HNO_3$$

and can be produced by dehydration of 100% nitric acid by means of phosphorus pentoxide, P_2O_5.

It is a white crystalline solid, volatile, and sublimes (passes directly into the vapor state from the solid, like iodine) at atmospheric pressure at 32.4°C. The crystals are hygroscopic and dissolve readily in water to give nitric acid.

It is a strong oxidizing agent and reacts, sometimes violently, with metals and organic compounds.

Dinitrogen pentoxide has been of particular interest to physical chemists by affording, in its decomposition, a classic example of a monomolecular reaction. It does decompose spontaneously, slowly at low temperatures, more rapidly at higher, and the rate at which it disappears according to the over-all reaction

$$2N_2O_5 = 4NO_2 + O_2$$

is determined, at any one temperature, only by the concentration of N_2O_5 present at the time

$$-\frac{dp_{N_2O_5}}{dt} = kp_{N_2O_5}.$$

The knowledge of calculus required to deal with this rate equation is pretty elementary, and the result of the integration is usually expressed in giving what is called a "half-life" value. You may recall reading that radioactive substances, like radium and cobalt-60, decompose according to this same relationship. The half-life is the time required for half of the atoms of a radioactive substance present at a given time to undergo radioactive decay. For N_2O_5, half-life values vary with temperature as follows:

t, °C	$\tau_{\frac{1}{2}}$
25	6 hr
35	86 min
100	5 sec

A great deal of effort, as you can imagine, has gone into the study of this reaction, since it is hard to understand what would make it just decompose all by itself. The mechanism proposed by Ogg[10] seems to be the most satisfying and at the same time the most helpful in understanding the reactions of the other oxides. He invokes the next higher oxide, NO_3, which we might as well anticipate right now, and postulates an equilibrium between it and N_2O_5:

$$N_2O_5 = NO_3 + NO_2.$$

But NO_3 is unstable and decomposes:

$$NO_3 = NO + O_2.$$

And NO reacts rapidly with NO_3:

$$NO_3 + NO = 2NO_2.$$

This last reaction is then the slow reaction, at the extremely low concentrations at which NO_3 is present, and is thus the rate-determining step in the sequence. But since the amount of NO_3 present is proportional to the concentration of N_2O_5, the rate is proportional to the concentration of N_2O_5; in other words, it behaves as a monomolecular reaction.

You have probably not seen a reference in your chemistry textbooks to this higher oxide, NO_3, called dinitrogen hexoxide, to distinguish it from N_2O_3, though there is no evidence that it exists in the dimeric form. There is sufficient evidence that it does exist, though it is highly unstable. An oxide with this composition can be produced by passing a mixture of oxygen and NO_2 at low pressure through an electrical discharge tube, and then

[10] R. A. Ogg, Jr., *Journal of Chemical Physics*, **15**, 337–338 (1947).

through a condenser (called a "trap") cooled by liquid air at about $-190°C$. It can also be made by the action of ozone on dinitrogen pentoxide, and characteristic absorption lines attributable to it can be observed spectroscopically. These lines are observed in the course of the decomposition of N_2O_5, thus supporting mechanism of this reaction just proposed.

Rate of Oxidation of Nitric Oxide

Having learned this much about the other oxides of nitrogen, we are now prepared to go back and examine in detail the reaction between nitric oxide and oxygen,

$$2NO + O_2 = 2NO_2,$$

which we discussed kinetically before, and which is of overriding importance in the manufacture of nitric acid. In view of the relatively infrequent occurrence of three-body collisions, it is natural to postulate some intermediate compound that could be formed by a more likely bimolecular reaction, which would then proceed to react with a third molecule. Some chemists have suggested N_2O_2, a dimeric form of NO, as this intermediate, but its existence in the gas phase has never been demonstrated. A more appealing mechanism was offered by Farrington Daniels and one of his co-workers.[11] They proposed that there is an initial reaction between oxygen and nitric oxide which produces a small, equilibrium concentration of NO_3:

$$NO + O_2 = NO_3.$$

The NO_3 then reacts rapidly, it is supposed, with the excess of NO to produce NO_2:

$$NO + NO_3 = N_2O_4 = 2NO_2$$

It seems certain that the equilibrium of the first reaction would be more favorable to the formation of NO_3 at low temperatures than at higher, and the low concentrations producible at ordinary temperatures would account for the relatively slow rate of the

[11] J. C. Treacy and F. Daniels, *Journal of the American Chemical Society*, **77**, 2033 (1955).

over-all reaction. The rate of the second reaction would be proportional to the concentration of NO multiplied by that of NO_3. This latter, in turn, would be proportional to the concentration of NO multiplied by that of O_2. In effect, then, the rate of the over-all reaction would be proportional to the concentration of oxygen multiplied by that of the nitric oxide squared—just as observed.

While this line of thought provides a satisfactory hypothesis to account for the rate of this termolecular reaction, it is not the only one. Even the matter of three-body collisions ought not to be casually dismissed, as we have done. Hinshelwood,[12] by assigning fairly reasonable values for the molecular dimensions, could come close to accounting for the temperature coefficient of the reaction rate, that is, to the "energy of activation" of the reaction. And Eyring's theory of the "activated complex," a transistory intermediate aggregation, not necessarily a compound that can be isolated, a theory you will be learning more about if you continue your studies in chemistry, yields values of the reaction rate in fairly close agreement with those observed, and a negative temperature coefficient, as required.[13] It is equally successful in predicting the rate of the reaction between nitric oxide and chlorine

$$2NO + Cl_2 = 2NOCl$$

which is termolecular also, and slow, but with a small positive temperature coefficient.

The reaction between NO and oxygen can be catalyzed, for example by silica gel. A great deal of work was done on this possibility in connection with the "Wisconsin process," a thermal process for the fixation of atmospheric nitrogen, to be discussed briefly later. But it is still rather slow, and the silica gel is inactivated by water vapor. So industrially, the rate is increased most economically in the ammonia oxidation process by increasing the total pressure.

[12] C. N. Hinshelwood and T. E. Green, *Journal of the Chemical Society* (*London*), **1926**, 730.

[13] S. Glasstone, K. J. Laidler, and H. Eyring, *The Theory of Rate Processes*, pp. 275–279, McGraw-Hill, New York, 1941.

A point worth noting in regard to the chemistry of the nitrogen oxides is that, with the possible exception of nitrous oxide, they are all interconvertible in the presence of oxygen. Indeed, they are generally in equilibrium with one another, the direction of the equilibrium being dependent of course on the temperature. This becomes more apparent when we consider their reactions with water to produce oxyacids of nitrogen, which we will now proceed to do.

There are two oxyacids of nitrogen of importance to us, nitrous acid, HNO_2, and nitric acid itself, HNO_3.

Nitrous Acid

Nitrous acid has already been mentioned in connection with dinitrogen trioxide N_2O_3, which is its anhydride. Nitrous acid is, indeed, formed in water solution, when gases corresponding to this composition react with water, but it is not the sole product, on account of its instability. A water solution of nitrous acid can be readily prepared by mixing a solution of barium nitrite with dilute sulfuric acid in stoichiometric proportions; when the barium sulfate formed precipitates, it can be filtered off:

$$Ba(NO_2)_2 + H_2SO_4 = BaSO_4 + 2HNO_2.$$

But the HNO_2 decomposes, slowly when cold, more rapidly when heated, evolving NO:

$$3HNO_2 = HNO_3 + H_2O + 2NO.$$

Nitrous acid in the vapor state exists in equilibrium with its decomposition products, and the equilibrium has been determined, conveniently enough, by photometric measurement of the NO_2 concentration. For the forward reaction in the gas phase, we may write

$$NO + NO_2 + H_2O = 2HNO_2$$

and set up an equilibrium constant

$$K_p = \frac{(p_{HNO_2})^2}{(p_{NO})(p_{NO_2})(p_{H_2O})}$$

Values of this equilibrium constant were found as follows:[14]

t, °C	K_p
20	1.56
40	0.641
60.5	0.250
80.6	0.112

It is evident that the HNO_2 vapor is more completely decomposed at higher temperatures. Of course, account must be taken, in making such determinations, of the simultaneous equilibria established between NO and NO_2 in forming N_2O_3, and the association of NO_2 to form N_2O_4, and also the reaction between NO_2 and water to form HNO_3, which we shall examine presently. To show how these are involved, the following data are quoted from the same paper, giving the partial pressure of each of these constituents at equilibrium at two temperatures, all being in the vapor phase.

t, °C	Partial pressure, mm Hg						
	NO_2	N_2O_4	N_2O_3	NO	H_2O	HNO_2	HNO_3
20.0	15.5	3.4	8.00	510.5	12.9	13.7	0.041
80.7	17.7	0.05	0.44	440.8	98.9	10.65	0.039

Nitrous acid is a weak acid, with an ionization constant at 25°C of 4×10^{-4}, a little higher than that of formic acid (1.76×10^{-4}) or acetic acid (1.75×10^{-5}).

The decomposition of nitrous acid is accelerated by hydrogen ions. When neutralized by alkali, the nitrite ion is stable enough, and nitrite salts of most bases can be prepared and are stable up to moderate temperatures. With ammonia, however, or with urea, nitrous acid reacts to give molecular nitrogen. With primary aromatic amines it reacts to give diazonium salts that are useful intermediates in the manufacture of dyestuffs:

$$RNH_2 + HNO_2 + HCl = RN{\cdot}NCl + 2H_2O.$$

[14] P. G. Ashmore and B. J. Tyler, *Journal of the Chemical Society (London)*, **1961**, 1017.

With secondary amines, nitroso compounds result:

$$R_2NH + HNO_2 = R_2N \cdot NO + H_2O.$$

Nitrous acid can act both as an oxidizing and a reducing agent. It oxidizes iodide ion and thiosulfate ion, but is itself oxidized by permanganate. This latter reaction is used for its quantitative determination. It can also be oxidized to nitric acid by hydrogen peroxide.

The reaction of gases containing NO and NO_2 with water to give nitrous acid is of technical interest, but it has not been studied as extensively as the reaction that forms nitric acid, and is best looked at in connection with that reaction, since some nitric acid is always formed.

With alkaline solutions, however, such gases react, if NO is in stoichiometric excess, as N_2O_3 to the extent that NO_2 is present, and form nitrites. The excess of NO is not absorbed. Relatively pure sodium nitrite, substantially free from nitrate, can be produced commercially this way, by bringing gases containing nitrogen oxides, in a low state of oxidation, into contact with a solution of caustic soda, NaOH, or of sodium carbonate, Na_2CO_3 ("soda ash"), if care is taken to keep the solution always on the alkaline side.

Chemistry of Nitric Acid

We come finally to nitric acid and to the reaction by which it is produced from nitrogen peroxide and water, the last step in making nitric acid by ammonia oxidation.

You have doubtless used nitric acid as a laboratory reagent, from a stock-room bottle labeled "Concentrated HNO_3." Actually, this is a water solution, free from impurities to the extent shown on the label, but containing only about 68% HNO_3 by weight. This concentration corresponds to the constant-boiling mixture that HNO_3 forms with water, which has the composition at atmospheric pressure 68.4% by weight, boiling at 121.9°C.

Pure, that is, 100% HNO_3 can be prepared by the method

described in Chapter One, or by distillation of a water solution with concentrated sulfuric acid. Nitric acid is a colorless liquid, with a specific gravity of 1.503 at 25°C, a freezing point of −41.6°C, and a boiling point of 86°C. The 100% acid is not entirely stable, however. The rate of decomposition is slow at low temperatures and in the dark. On exposure to light, or on heating, it decomposes more rapidly. Even at the boiling point, or just above, a matter of days is required for the vapor to come close to the equilibrium in the reaction

$$2HNO_3 = H_2O + 2NO_2 + \tfrac{1}{2}O_2.$$

The equilibrium constant for this reaction,

$$K_p = \frac{(p_{H_2O})(p_{NO_2})^2(p_{O_2})^{\frac{1}{2}}}{(p_{HNO_3})^2},$$

at 25°C is 6.9×10^{-5} (pressures in atmospheres), so you see the extent of decomposition is low at ordinary temperatures.

The heat of formation of NHO_3 in the gaseous state, at 25°C, is 32,000 cal/mole (exothermic, $\Delta H = -32{,}000$). The heat of vaporization (or condensation) is 9420 cal/mole; the heat of solution to give an acid of 63% by weight is about 2540 cal/mole HNO_3. Taking all these factors into account, and remembering that the heat of formation of NO is 20,650 cal/mole, endothermic ($\Delta H = 20{,}650$), we can get a figure for the total amount of heat we have to consider in the second and third steps in the ammonia oxidation process. Let us suppose we bring all the primary products of the initial oxidation of ammonia to 25°C without condensing any water or allowing any nitric oxide to oxidize to NO_2, and convert them all to nitric acid of 63% by weight. Writing the equation as

$$2NO + \tfrac{3}{2}O_2 + H_2O(g) = 2HNO_3(aq)$$

and putting in the heats of formation (ΔH_f) as just stated, with a tabulated value for H_2O,

$$2(20{,}650) + O + (-58{,}300) = 2(-45{,}960),$$

we get

$$\Delta H = -74{,}900$$

as the heat of reaction to produce 2 moles of HNO_3, or 37,450 cal/g mole heat evolved in the sequence of reactions, as you can check for yourself. Now this is 595 cal/g, which is the same as 595 pound-centigrade units (pcu) per pound, and since the centigrade degree equals 1.8 Fahrenheit degrees, this amounts to 1170 Btu (British thermal units, or pound-Fahrenheit units) per pound of HNO_3 produced. This may not sound like much to you, but when you read later on that a commercial plant may make 240 tons of HNO_3 per day, or 10 tons, 20,000 lb/hr, you will realize that the amount of heat to be dissipated in the absorption system of such a large plant is 23,400,000 Btu/hr.

To get a more vivid feeling for the amount of heat involved, let us calculate how much cooling water this would take. One gallon of water weighs 8.33 lb, and 1 gal/min equals 500 lb/hr. The specific heat of water is 1.0; this is the definition of the Btu, namely, the heat required to raise the temperature of 1 lb of water 1°F. If we can reasonably heat cooling water 50°F, from say 75°F to 125°F, then 1 gal/min will take up 25,000 Btu/hr, and it will take 935 gal/min to absorb the 23,400,000 Btu/hr given off in the absorption system of a large nitric acid plant. You have here another example of the kind of preliminary design calculation the chemical engineer has to make in proposing to carry out a process commercially.

Nitric acid is miscible with water in all proportions. It forms hydrates at low temperatures: $HNO_3 \cdot H_2O$, melting at −37.68°C, and $HNO_3 \cdot 3H_2O$, melting at −18.47°C. Eutectics between these hydrates and between them and water, on the one hand, and pure HNO_3 on the other, are shown in the melting point–composition diagram (see Figure 2.2).

Evidence of incipient compound formation is also found in the vapor–liquid equilibrium relationships. The addition of HNO_3 to water increases the boiling point until the maximum of 121.9°C is reached, at 68.4% by weight (38.3 mole per cent). Conversely, the addition of water to pure HNO_3 increases the

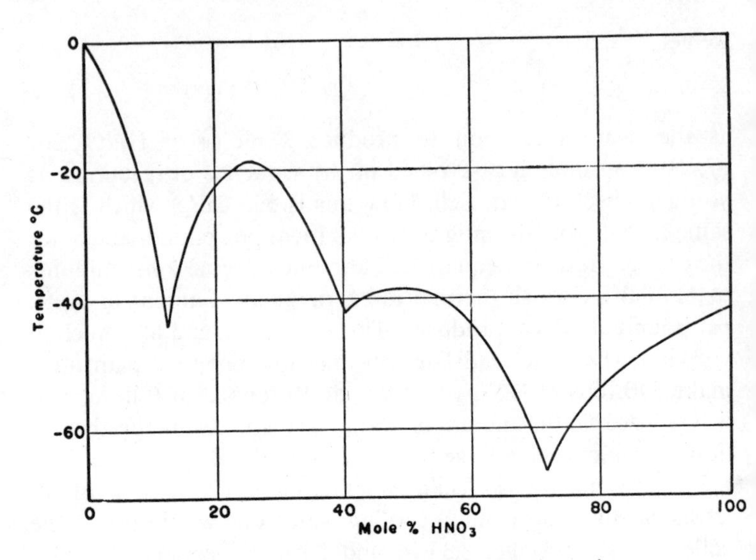

Figure 2.2 The system H_2O—HNO_3, freezing points.

From M. C. Sneed and R. C. Brasted, *Comprehensive Inorganic Chemistry*, Vol. 5, p. 92, Van Nostrand, New York, 1959, with permission.

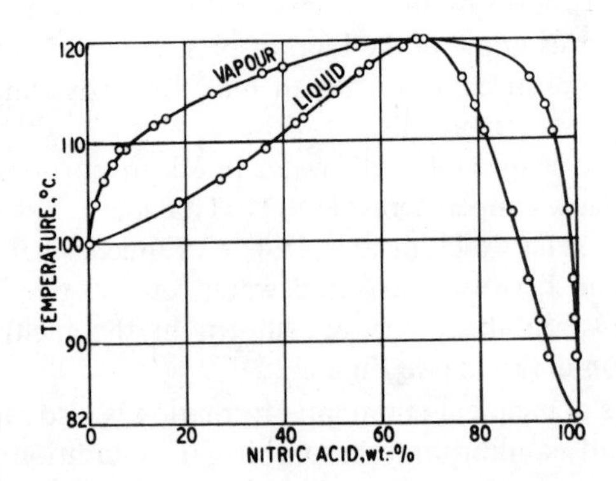

Figure 2.3 Nitric acid—water system, vapor-liquid compositions.

From S. R. M. Ellis and J. M. Thwaites, "Nitric Acid—Water—Sulphuric Acid," *Journal of Applied Chemistry (London)*, 7, 154, 157, 1957, with permission.

boiling point from 86°C to this same maximum value, as shown in the vapor–liquid composition diagram[15] (see Figure 2.3).

It is this relationship that makes it necessary to employ some dehydrating agent to get high-strength nitric acid (95% or more), as required in many applications, when using the ammonia oxidation process that produces acid of a strength below 68% by weight, when operated at atmospheric pressure. (See Chapter Six.)

Paradoxically, nitric acid behaves as a "strong" acid in dilute solution; that is, it is ionized almost completely to hydrogen H^+ and nitrate NO_3^- ions, while in concentrated solutions nitronium ions and nitrate ions are produced from two molecules:

$$2HNO_3 = H_2O_3N^+ + NO_3^-;$$

$H_2O_3N^+$ may be looked on as

$$(NO_2 \cdot H_2O)^+.$$

In this sense HNO_3 would parallel the decomposition of its anhydride, N_2O_5, into $NO_2 + NO_3$; N_2O_5 can be looked on as nitronium nitrate. With other strong acids, HNO_3 reacts partly as a base, for example, with perchloric acid; at any rate N_2O_5 does:

$$N_2O_5 + HClO_4 \rightarrow \underset{\text{nitronium perchlorate}}{(NO_2)ClO_4} + HNO_3.$$

Nitronium perchlorate is a crystalline compound,[16] decomposing without melting at a temperature of 135°C or above, and reacting violently with many organic compounds. It has been considered as alternative for ammonium perchlorate used as an oxidant in solid-fuel rocket propellant mixtures.

With organic compounds, nitric acid may act (*a*) simply as an acid, (*b*) as an oxidizing agent, or (*c*) as a nitrating agent. An

[15] S. R. M. Ellis and J. M. Thwaites, *Journal of Applied Chemistry* (*London*), **7**, 152–160 (1957).

[16] D. R. Goddard, E. D. Hughes, and C. K. Ingold, *Journal of the Chemical Society* (*London*), **1950**, 2559–2575.

example of (*a*) is the formation of glyceryl trinitrate ("nitroglycerin") by esterification, to use the organic chemistry term, in the presence of sulfuric acid:

$$C_3H_5(OH)_3 + 3HNO_3 = C_3H_5(NO_3)_3 + 3H_2O.$$

As an example of (*b*), we may cite one of several reaction paths in the oxidation of cyclohexanol (derived from benzene or from cyclohexane from petroleum) to produce adipic acid, an intermediate for the manufacture of nylon:

$$C_6H_{11}OH + 8HNO_3 = (CH_2)_4(COOH)_2 + 8NO_2 + 5H_2O.$$

For (*c*), the classic example is the nitration of benzene to give nitrobenzene, which can then be "reduced" to aniline:

$$HNO_3 + C_6H_6 = C_6H_5NO_2 + H_2O.$$

This reaction, like the first, also requires the presence of concentrated sulfuric acid. (See Chapter Seven.)

Nitric acid dissolves metals, not only iron but also copper and silver, though generally with the liberation of lower oxides of nitrogen rather than hydrogen. The action on metals is more rapid initially if some lower oxides are already present. The addition of HCl gives *aqua regia*, capable of dissolving the noble metals, gold and platinum. This result is not due to any higher oxidation potential of the acid mixture but rather to the high concentration of chloride ions that increase the solution potential of the metal because of the tendency to form chloride complexes.

One peculiarity of behavior of nitric acid is of industrial importance, since the modern process for making the acid depends on it. This is the phenomenon known as passivity. It has long been observed that a clean piece of iron immersed in strong nitric acid at room temperature does not dissolve; it is said to become "passive." The same property was shown to an even greater degree by chromium and chromium–nickel alloys of iron, developed in England and in Germany, respectively, beginning in 1913. These alloys, introduced first in cutlery, are known as "stainless steel." Suitably formulated alloys, notably 15 to 16%

Cr, and 18% Cr–8% Ni (popularly called "18–8"), are usefully resistant to nitric acid of all concentrations (at least up to 95% by weight) and at temperatures up to the atmospheric pressure boiling point. Their resistance to attack is attributed to the formation of a protective oxide layer on the surface of the metal.

High-strength nitric acid, above 95% by weight, is resisted better by aluminum (preferably the alloy now known as "3003," containing 1.2% of manganese, balance aluminum), but aluminum should not be exposed for any length of time to acid of lower strengths.

Titanium is found useful, especially in the presence of organic acids.

Stainless steel is attacked by nitric acid in the presence of chlorides (as is aluminum). It is thought this is because the chloride ions can penetrate the protective oxide layer. For this reason, stainless alloys would not have been serviceable in producing nitric acid from Chilean nitrate, which always contained chlorides. They were, however, adaptable to the ammonia oxidation process, which gave an acid substantially free from impurities.

Let us now take up the reaction by which nitric acid is produced from nitrogen oxides and water. This reaction can be represented by the equation

$$3NO_2 + H_2O = 2HNO_3 + NO$$

for the over-all process. The reaction has been extensively studied, in view of its recognized industrial importance, and not all points have been cleared up yet. Both equilibrium and kinetic factors are involved, and the mechanism of the reaction, the successive steps by which it proceeds, has not been conclusively established.

As it occurs in practice, near room temperature, liquid water is involved, and the HNO_3 is in water solution, while the nitrogen oxides are present in the gas phase. There is thus a heterogeneous equilibrium between the phases for each component, as well as the homogeneous equilibrium in the gas phase toward which the concentrations in that phase will tend. Measurements of the equilibrium constant for this homogeneous reaction, set up as

$$K_p = \frac{(p_{HNO_3})^2(p_{NO})}{(p_{NO_2})^3(p_{H_2O})}$$

have been made, and the results of different investigators are in pretty close agreement, representative values at 25°C being 0.015 and 0.019. Values calculated from thermochemical data likewise are in agreement, as to order of magnitude (giving 0.0105 at 25°C, 298°K), and show the following variation with temperature.

t, °K	K_p
275	0.0375
300	0.0095
350	0.0010
400	0.0002

Here again, as with most of the other reactions involved in the process, you can see that completion of the reaction in the direction of formation of nitric acid is favored by low temperatures.

When the liquid phase is present, as in the actual case, the partial pressures of water and of nitric acid in the vapor phase are fixed by the concentration of HNO_3 in solution, and a useful relationship can be set up by substituting the known partial pressures at corresponding temperatures in the equilibrium expression, obtaining values of K_1

$$K_1 = \frac{(p_{NO})}{(p_{NO_2})^3}$$

as a function of acid concentration, at various temperatures, as represented in Figure 2.4.

The uppermost curve represents the lowest temperature, 10°C, and it is again apparent that for any one acid concentration how much more favorable (to the formation of HNO_3) it is to keep the temperature low. It is also apparent that at any one temperature, completion of the reaction as written is favored by keeping the acid concentration low. But our industrial objective is to get as strong an acid as possible. That means we have to get as high a

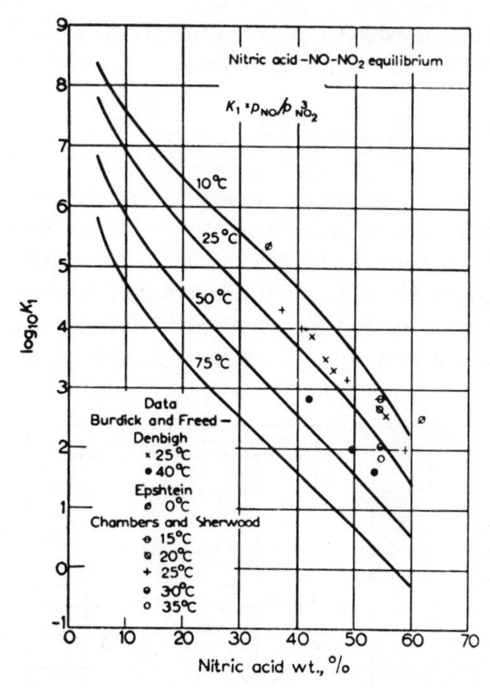

Figure 2.4 Nitric acid—NO—NO_2 equilibrium.

After J. J. Carberry, *Chemical Engineering Science*, *9*, 190 (1959), with permission.

ratio of NO_2 to NO as possible, and keep the temperature low as well.

Now you will remember that there are other equilibria involved as well: NO_2 with N_2O_4, and NO_2 and NO with N_2O_3, and these gases with HNO_2 vapor. All of these equilibria are rapidly attained. There are, of course, other gases present. Oxygen is present, but the reactions into which it enters, either with NO on the one hand or with NO_2 and H_2O on the other, are slow, and we shall consider them separately. Nitrogen gas is present, but it does not enter into these reactions; it is just as inert as the argon contained in the atmosphere.

The equilibrium between NO_2 and N_2O_4 does come into play, and can be taken into account. We can substitute for

(p_{NO_2}) in the expression for K_1 the value we get from that for K_3, on p. 39, in terms of $(p_{N_2O_4})$, and obtain

$$K_4 = \frac{(p_{NO})}{(p_{N_2O_4})^{1.5}} = \frac{K_1}{(K_3)^{1.5}}.$$

Substituting actual values, we get[17] the curve represented by Figure 2.5. Curiously enough, all the values lie on this single curve, and the temperature effect seems to be entirely taken care of by the variation of K_3 with temperature. But you will recall that association to N_2O_4 is favored by low temperature, and higher concentrations of N_2O_4 will drive the equilibrium toward completion of the reaction.

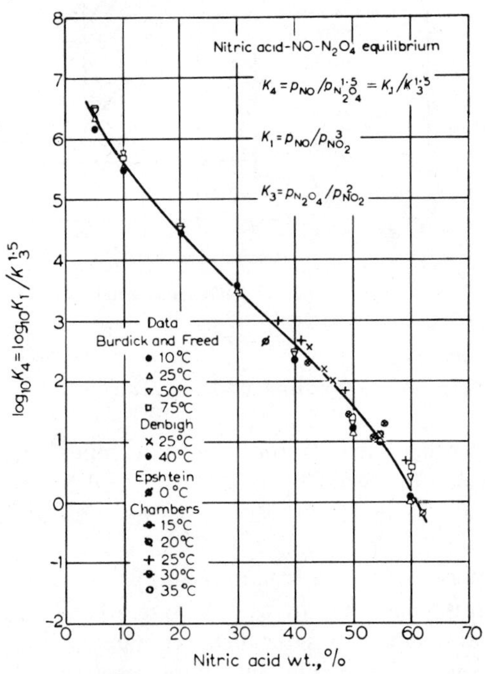

Figure 2.5 Nitric acid—NO—N_2O_4 equilibrium.

After Carberry, *op. cit.*, p. 191, with permission.

[17] J. J. Carberry, *Chemical Engineering Science*, **9**, 189 (1959).

So much for the equilibrium factors for this over-all reaction. What about the kinetic factors? They are also important in analyzing or predicting the performance of an absorption system designed to produce nitric acid from nitrogen oxides and water.

As far as the reaction can be carried out as a homogeneous gas-phase process, there is evidence that it is rapid. But the vapor pressure of the nitric acid solutions we want to produce is so low at the temperatures favorable for a high degree of completion of the reaction that we inevitably have to deal with reactions involving the liquid phase. We are then confronted with questions concerning the rate of transfer of material between the phases, which can occur only at the interfacial surface. There may also be slow reactions in the liquid phase. As it turns out, all of these points are involved.

In the first place, some nitrous acid is always found along with the nitric acid in solution, especially in dilute solutions. In addition some dissolved and unreacted NO_2 is present in acid of higher concentrations, as evidenced by its brown color. The mechanism by which the reaction proceeds must be as follows: the dioxide reacts in water solution to produce nitric and nitrous acids,

$$6NO_2 + 3H_2O = 3HNO_3 + 3HNO_2.$$

But nitrous acid is unstable in the presence of stronger acids and decomposes,

$$3HNO_2 = HNO_3 + 2NO + H_2O.$$

The net reaction is thus,

$$6NO_2 + 2H_2O = 4HNO_3 + 2NO,$$

or

$$3NO_2 + H_2O = 2HNO_3 + NO,$$

as given previously. In a continuous operation the nitrous acid builds up to a steady concentration at which it decomposes as fast as formed and so does not offer any real hindrance to the process, and the equilibrium we were looking at tends to prevail as written.

Naturally, there are rate factors involved in the process of dissolving the nitrogen peroxide, and some investigators have attempted to show that the rate of absorption from the gas phase into the solution follows the same laws as for the solution of other soluble gases. To a certain extent they have succeeded. The rate of absorption is proportional to the interfacial surface area exposed, and is influenced to some extent by the conditions of turbulence in the gas and in the liquid phase. But the rate of absorption in pure water (where the acid concentration is zero, and the equilibrium constant has such a high value that the reverse reaction does not need to be taken into account) is found to be directly proportional to the concentration of dinitrogen tetroxide, N_2O_4, in the gas phase, as given by the equilibrium constant for the reaction

$$2NO_2 = N_2O_4$$

under the conditions prevailing, and not to the concentration of the dioxide, NO_2. Further, the latest and most careful work appears to show that the slowest, and therefore the controlling, step in the sequence is the net rate of ionization of the dissolved peroxide (as tetroxide),

$$N_2O_4 + H_2O = HONO + H^+ + NO_3^-.$$

This finding leads back once more to the conclusion that the production of nitric acid is favored by a high concentration of N_2O_4, and this in turn requires a high concentration of peroxide and a low temperature; so the technological implications of these investigations are clear.

Now we have to return to the reaction between oxygen and nitric oxide and we see that for every two molecules of nitric acid we get one molecule of nitric oxide which requires reoxidation to the peroxide before it can again be reacted with water. As pointed out before, this is the really slow stage in the operation, and accounts for the large volume that must be provided in an absorption system for making nitric acid. The considerations we looked at before prevail, and lead to the same conditions favoring the reaction: high concentrations of the reactants and low temperatures.

We now conclude our examination of the steps by which oxides of nitrogen produced by burning ammonia are converted to nitric acid. But this discussion of the properties of nitric acid cannot be terminated without a reference to safety precautions.

Safety Precautions

Nitric acid is a dangerous chemical, and is to be handled with due regard to its corrosive and oxidizing properties.

A dilute solution is harmless enough. It used even to be prescribed as a tonic. In higher concentrations it would be poisonous. There is a truism to the effect that "a poison is too much" – of anything. A person simply could not swallow strong nitric acid deliberately; but it would not take very much to cause fatal damage to his internal organs if he did so accidentally, for example, by careless manipulation of a pipette. (Caution! Use a suction bulb to fill a pipette, never the mouth.) Spilled on the skin, it causes yellow discoloration at the least, and larger quantities can cause serious and even fatal burns.

Avoid contact with the acid, and put on protective equipment, most particularly, eye protection (safety goggles with side shields, or better, cup-type rubber-framed goggles) if you have to handle it in any quantity. In your place of work there should be a copy of "Chemical Safety Data Sheet SD–5, Nitric Acid," put out by the Manufacturing Chemists' Association. A current edition is available for 30¢ from the Association, 1825 Connecticut Avenue, N.W., Washington, D. C.

Do not breathe the fumes that are given off when nitric acid dissolves metals or oxidizes organic compounds. They are toxic, and you can suffer delayed serious effects without noticing immediate discomfort. Stay out of any red-brown fumes!

Nitric acid does not itself take fire, but it oxidizes organic matter, wood, cloth, or paper, to a form that is highly flammable. Any spills should therefore be flushed copiously with water, and any wood or other cellulosic product exposed to it should be disposed of immediately, and with caution. Nitric acid is still *aqua fortis*, "strong water."

The decomposition of nitric oxide in the liquid state (or solid suspended in liquid) has been found to be violently explosive, giving rise to a shattering detonation when subjected to shock. It should therefore be handled with caution. (R. O. Miller, *Industrial and Engineering Chemistry, Process Design and Development*, **7**, no. 4, 590–592 (1968).)

CHAPTER THREE

AMMONIA OXIDATION; ATMOSPHERIC PRESSURE PROCESS

Nitric acid is produced nowadays exclusively by oxidation of ammonia. How did this process originate and how is it accomplished?

The fixation of atmospheric nitrogen – the transformation of the inexhaustible supply of free, gaseous nitrogen in the air into a form more readily utilized by plants in the processes by which they produce the proteins required for their own growth and for the growth of animals (including man)—was one of the most challenging technological problems of recent times. Dire predictions were made around the turn of the century (1900) as to the future of the human race, or at any rate of the Western nations accustomed to a diet of wheaten bread, unless a solution was found.

Organic wastes had of course been employed from the dawn of civilization, to renew and increase the fertility of cultivated fields. The early American colonists found that the Indians were aware of the "magical power" of a dead fish put into every hill of corn they planted. While after a fashion such sources sufficed in primitive or agricultural eras, they were by no means enough to support the population as it began to increase during the Industrial Revolution. You read in Chapter One how nitrate of soda from Chile came into prominence as a fertilizer material, but its inevitable depletion could be predicted.

Faced with such a challenge, the ingenuity of chemists in every country was devoted to a study of any and all reactions which elemental nitrogen could be made to undergo, in the hope that one of them might be brought to economic realization.

Processes for Nitrogen Fixation

Naturally, one of the first reactions to be studied was the direct union of nitrogen with the oxygen of the air, under the influence of an electric spark. This reaction occurs in nature during lightning flashes, and furnishes some "fixed" nitrogen to the soil in rain water. It was first observed in the laboratory by Joseph Priestley in England about 1775, and studied by Cavendish, who reported his experiments in 1785.[1]

With the introduction of dynamoelectric machines a century later (following the invention of the electric light), this process for nitrogen fixation began to appear attractive. It was, indeed, extensively studied, and practiced commercially. It produces NO, nitric oxide, directly, which can be converted to nitric acid. Thus, constituting an alternative process to ammonia oxidation, it will be dealt with at some length in Chapter Five.

What other reactions will elemental nitrogen gas undergo? All appeared to require high temperature, a natural inference from its recognized inertness under ordinary conditions. It was found early in the nineteenth century that nitrogen would combine at a red heat with carbon and potash to produce potassium cyanide:

$$4C + K_2CO_3 + N_2 = 2KCN + 3CO.$$

The cyanide could be hydrolyzed with water or with acid to produce ammonia or ammonium salts. This reaction was tried in various countries, sometimes employing calcium or barium instead of potassium. One such attempt was made in the United States during World War I, to exploit the so-called Bucher process for sodium cyanide, but like all the others it failed to achieve commercial success.

A process discovered about 1898 by Frank and Caro in Germany, and which has survived (though not now used for ammonia

[1] *Philosophical Transactions of the Royal Society of London*, **75**, 372 (1785).

production), is that based on the reaction between calcium carbide and gaseous nitrogen, again at a high temperature:

$$CaC_2+N_2 = CaCN_2+C.$$

(Calcium cyanamide, $CaCN_2$, is the calcium derivative of $N{=}C{-}NH_2$, the amide of hydrocyanic acid $N{=}C{-}H$.) Calcium carbide had become available as the result of the development of electric furnace processes that sprang up in the last few years of the nineteenth century, centered around sites where hydroelectric power was being developed. The process soon spread through Europe and was introduced into North America early in the 1900s, and became the primary material produced by the American Cyanamid Company. Calcium cyanamide can be hydrolyzed to give ammonia:

$$CaCN_2+3H_2O = CaCO_3+2NH_3.$$

This was the only established nitrogen fixation process in America at the time of World War I. It was on that account selected as the basis for the enormous (for that time) plant built by the U.S. Government at Muscle Shoals, Alabama, known as U.S. Nitrate Plant No. 2.

Calcium cyanamide still finds limited application as a fertilizer (chiefly in Europe). It serves usefully as a starting point for the manufacture of derivatives such as guanadine and melamine. The latter is the raw material for the synthetic resins that you see under the name of "Beetle Ware." Cyanamide no longer competes in the production of ammonia.

One more unsuccessful process might be mentioned in passing. At high enough temperatures nitrogen combines with reactive metals, for example, titanium. There was talk at one time of basing a process on the coproduction of titanium nitride in the reduction of titaniferous iron ores. A project that got beyond the talking stage was the so-called Serpek process, in which the aluminum ore, bauxite, was reacted with carbon at 1600°C in the presence of nitrogen:

$$Al_2O_3+3C+N_2 = 2AlN+3CO.$$

The aluminium nitride was then hydrolyzed with water to produce aluminum hydroxide (for feed to aluminum reduction cells) and ammonia:

$$2AlN + 6H_2O = 2NH_3 + 2Al(OH)_3.$$

Though tried on a fairly large scale by a French concern, it was abandoned for lack of suitable refractories to withstand the reaction conditions.

Ammonia Synthesis

With that cursory review of unsuccessful processes for the fixation of nitrogen in the reduced form, as ammonia, let us direct our attention briefly to the process that has emerged as successful and which is now practiced all over the world; namely, the direct synthesis of ammonia from nitrogen and hydrogen. This is a development of such outstanding significance in modern technology that it deserves to have a book written about it exclusively. If you are interested, look for further information into the American Chemical Society Monograph No. 161,[2] a source from which I have drawn in what follows.

Looking backward fifty or sixty years, it is hard to understand how chemists were so slow to realize the possibility of catalysis and of high pressure in causing nitrogen and hydrogen to form ammonia. Le Châtelier's principle, interpreted for this case as indicating that a reaction which involves a decrease in volume, such as

$$3H_2 + N_2 = 2NH_3,$$

would be favored by high pressure, had been enunciated some years before.

The Law of Mass Action was established, and Nernst was working out his "heat theorem," from which equilibrium conditions could be estimated from thermochemical values. This possibility is now familiar to students of physical chemistry in the relation of the "free energy change" involved in a reaction to the

[2] V. Sauchelli, *Fertilizer Nitrogen: Its Chemistry and Technology*, American Chemical Society Monograph No. 161, Reinhold, New York, 1964.

equilibrium constant. But the possibilities of catalysis had not then been exploited, and techniques for handling gases at higher than ordinary pressures and temperatures had not been developed. In his lectures at Karlsruhe in 1905, Professor Haber, who the year before had indeed produced a measurable quantity of ammonia from nitrogen and hydrogen (at 1 atm and 1000 C, over an iron wire), expressed the opinion that the commercial synthesis of ammonia was a long way off.

But it was this same Professor Haber who in 1908, with an English (not French) assistant named Le Rossignol, showed that increased proportions of ammonia could be obtained at 30 atm pressure.

In 1909 he was able to demonstrate to Carl Bosch of the large German chemical firm generally known by its initials, B.A.S.F. (for Badische Anilin- und Soda-Fabrik), the production of ammonia at a rate of 80 g/hr under a pressure of 185 atm, using a catalyst he had discovered, made from osmium oxide. He conceived the idea that efficient production could be achieved by condensing out the equilibrium concentration of ammonia and recirculating the unreacted gases over the catalyst.

The resources of the B.A.S.F. were thereupon thrown into the development of the process.

In the course of some twenty thousand experiments a more practical catalyst was developed, made from iron, with slight additions of alkali and alkaline earth oxides; this catalyst has been used without much change up to the present day. The technological difficulties in the preparation and purification of hydrogen and in the construction of suitable reactors were formidable but nonetheless were overcome. By 1913 a plant at Ludwigshafen-Oppau was producing fixed nitrogen at a rate of 30 tons/day.

During World War I, with supplies of Chile nitrate cut off by the naval blockade, Germany became desperate for nitric acid, as the stockpile accumulated in anticipation of a short war began to run out. Enlargement of the synthetic nitrogen plants was therefore hastened, and by the war's end, production capacity had reached 200,000 tons per year.

Haber was awarded the Nobel Prize in 1918 for his work on ammonia synthesis, and Bosch received the Nobel Prize in 1931 for his contributions in the same field.

In the United States only experimental work was completed during World War I, and no significant production was achieved. In the years following, worldwide exploitation of the method, with some variations, has occurred, and a recent figure shows production rates running over 10 million tons annually.

The consequent effect on competitive prices (in the United States) is shown in Table 3.1. These are prices in dollars for one "unit" of nitrogen content (that is, 20 lb, or 1% of a ton of 2000 lb). If you compare these figures with those showing the purchasing

TABLE 3.1
Average Wholesale Prices of Various Nitrogen Carriers

Calendar Year	Ammonium sulfate, $	Anhydrous ammonia, $	Calcium cyanamide, $	Sodium nitrate, $	Ammoniating Solutions, $	Urea	Natural organics, $
1900	2.79	—	—	2.37	—	—	2.57
1905	3.01	—	—	2.97	—	—	2.88
1910	2.64	—	3.43	2.76	—	—	3.63
1915	3.09	—	2.54	3.04	—	—	3.54
1920	4.08	—	3.40	4.44	—	—	8.71
1925	2.65	—	2.20	3.28	—	—	4.88
1926	2.52	1.75	2.19	3.27	—	—	4.62
1928	2.27	1.54	2.01	2.88	—	—	6.13
1930	1.79	1.40	1.65	2.49	—	—	4.50
1932	1.02	1.34	1.00	1.86	—	—	1.83
1935	1.13	1.09	1.20	1.47	1.07	—	3.38
1937	1.32	1.09	1.26	1.64	1.17	1.52	4.49
1940	1.37	1.09	1.20	1.68	1.22	1.32	3.55
1945	1.42	.72	1.44	1.75	1.07	1.37	4.85
1947	1.60	.72	1.98	2.50	1.03	1.39	9.71
1949	2.29	.94	2.82	3.15	1.23	—	8.45
1950	1.95	.91	2.26	3.00	1.20	2.02	8.66
1952	2.09	.97	3.11	3.34	1.20	3.03	9.23
1957	1.75	.99	2.62	2.75	1.21	2.43	8.15
1959	1.67	1.06	2.71	2.66	1.21	2.30	8.15
1960	1.67	1.07	2.71	2.75	1.27	2.28	6.80
1961	1.67	1.10	2.71	2.75	1.31	1.89	7.22

Basis: Prices for 20 lb N (one unit) at producing points or ports in bulk car lots.

Computed largely from published quotations in the *Oil, Paint and Drug Reporter*. The earlier quotations for cyanamid, ammonia, solutions, and urea were supplied by the producers. These prices are for spot purchases. Contract prices are usually lower than those given here.

From V. Sauchelli, *Fertilizer Nitrogen: Its Chemistry and Technology*, American Chemical Society Monograph No. 161, p. 341, Reinhold, New York, 1964, with permission.

power of the dollar in terms of cost-of-living indexes, you will see that prices for "fixed" nitrogen, in the form of anhydrous ammonia, have been effectively reduced over the twenty-five year period, 1935–1960. Improvements in technology have more than made up for rising cost factors.

Does this brief recital cause you to speculate how many opportunities for innovation and for the exercise of ingenuity lie undiscovered, awaiting the alert chemist, the resourceful engineer, the determined manufacturer?

Ammonia can of course be obtained from animal products. It is traditionally stated in chemistry textbooks that "sal ammoniac" (ammonium chloride) was so called because it was produced originally from the urine of camels that brought travelers to the temple of Jupiter Ammon, in Libya in North Africa. There was such a temple, but the derivation may be fancy rather than fact. The Greek word *ammos* merely means "sand," and the salt was reported to be found in the sands—perhaps it was only ordinary sodium chloride. In any event, the urea in urine is converted by bacterial action into ammonia. Salt is also present, and, on heating the evaporated solution, ammonium chloride will sublime, leaving sodium carbonate in the residue.

Other animal products, such as bones, horns, and hides, yield ammonia on dry distillation. It is almost forgotten that an old name for ammonia is "spirits of hartshorn," in allusion to its production by this means.

Vegetable products also contain proteins, and dry distillation (carbonization) of wood or of coal (also of vegetable origin) yields some ammonia. Its recovery from by-product coke ovens, or even earlier from coal converted to illuminating gas, is an established process. Some ammonia was recovered as a solution in water (aqua ammonia) from which anhydrous ammonia could be distilled for use as a refrigerant in ice making and cold storage. Much, however, was (and is) absorbed in sulfuric acid, giving ammonium sulfate, which enters the market as a fertilizer material, in competition originally with nitrate of soda, and now with synthetic ammonia. The rate of production is determined by the demand for coke from the iron and steel industry.

Properties of Ammonia

At this point, we ought to review briefly the physical and chemical properties of ammonia as they are concerned in the process for its oxidation to produce nitric acid.

Ammonia at ordinary temperature and pressure is a colorless gas with a strong characteristic odor. While toxic, its odor is so pungent that there is ample warning before really dangerous concentrations are encountered. It can be liquified at moderate pressures as shown in Table 3.2.

TABLE 3.2
Temperature and Vapor Pressure for Liquefaction of Ammonia

Temperature		Vapor Pressure	
°C	°F	mm Hg	lb/sq in. abs
−33.4	−28.2	760	14.7
−20.0	−4.0	1427	27.5
0.0	32.0	3221	62.3
20.0	68.0	6429	124.1

The boiling points shown, together with its high heat of vaporization, 5580 cal/g mole, its low freezing point, −77.7°C, and its high critical temperature, 132.9°C, indicate its suitability as a refrigerating medium. It was the principal material used for this purpose until the introduction in the 1920s of the fluorinated hydrocarbons (Freons).

From the standpoint of our interest in its use as the raw material for oxidation to nitric acid, Table 3.2 shows how it can be readily stored at moderate pressures as a liquid and vaporized at moderate temperatures even up to reasonably high pressures. With refrigeration it is stored, even transported, at low pressure, practically down to atmospheric pressure.

Ammonia is very soluble in water. The solution in equilibrium with the gas at 1 atm pressure contains 28% NH_3 by weight (the aqua ammonia of commerce).

The chemical properties of most interest to us are its flammability and its tendency to decompose on heated surfaces. The limits of flammability of ammonia in air are from about 16% by volume (lower limit) to about 26% (upper limit) depending somewhat on conditions of ignition and direction of propagation. In oxygen the limits are 25% to 75%, respectively, at atmospheric pressure. Increase of pressure widens the limits. The ignition temperature is 920° to 1000°C.

Ammonia is decomposed into its elements when heated and brought into contact with most substances, especially iron. Ceramic materials free from iron are not very active; quartz is without effect. Aluminum surfaces (actually covered with aluminum oxide) are not active. Nickel is much less active than iron, as are stainless steels. These same remarks apply to mixtures of ammonia and air prepared for the oxidation process.

Ammonia solutions in water, it should be remembered, are corrosive to copper and copper alloys, brass and bronze, especially in the presence of air, and contact with these metals must be avoided in handling ammonia and ammonium salts.

There is one generally unrecognized hazard of which you should be made aware. With mercury, ammonia can form a sensitive explosive, and mercury should not be used in manometers for measuring the flow of ammonia gas.

Oxidation of Ammonia—Chemical Principles

The chemical property of ammonia of direct interest to us here is its reaction with oxygen to produce nitric oxide. Of primary importance is the heat effect involved.

From tables of thermochemical data we find that the heat of formation of ammonia from the elements is given as −11,000 cal/g mole, and that of water (vapor) as −57,800 cal/g mole. That of nitric oxide, as we saw in Chapter Two, is +21,600 cal/g mole.

Then for the reaction

$$4NH_3 + 5O_2 = 4NO + 6H_2O,$$

we calculate the heat effect by simple arithmetic:

$$4 \times 21{,}600 + 6 \times (-57{,}800) - 4 \times (-11{,}000),$$

or $-216{,}000$ cal/g mole for the reaction as written, or $-54{,}100$ cal/g mole NH_3. Any ammonia not converted to nitric oxide but going to elemental nitrogen will give off more heat:

$$4NH_3 + 3O_2 = 2N_2 + 6H_2O$$
$$6 \times (-57{,}800) - 4 \times (-11{,}000) = -302{,}800.$$

But considering only the first reaction, as we may, since under suitable conditions the fraction converted to nitric oxide may be as high as 99%, generally above 95% in any case, we can calculate the temperature rise of the mixture upon reaction. Assuming that the oxidation is done with air, we can easily estimate the temperature rise for any volume (or mole) fraction x of ammonia in the entering gas. For we can take the molar specific heat of the diluent nitrogen and of the residual oxygen and of the NO as 7.0 cal/(g mole) (°C), and that of water as averaging approximately 8.34 over the temperature range involved and obtain our expression for the theoretical temperature rise as

$$\frac{54{,}100x}{7.0(1-x)+8.34(1.5x)} = \frac{54{,}100x}{7+5.51x}.$$

For an ammonia-air mixture containing 10% by volume, we obtain for the temperature rise $5410/7.551 = 717$°C. Keep that figure in mind as we proceed to discuss the conditions for carrying out the oxidation of ammonia.

The following list, for reference, shows how the temperature rise, for complete conversion to nitric oxide, varies almost linearly with entering ammonia composition:

% NH_3 by Volume	Temperature Rise, °C
9.0	650
10.0	717
11.0	784

The first recorded account of the production of oxides of nitrogen from ammonia was that contributed by the Reverend Isaac Milner[3] to the Royal Society in 1789. His quaint account of his experiments is worth quoting verbatim.

[3] *Philosophical Transactions of the Royal Society of London*, **79**, 606–613 (1789).

> Almost immediately on seeing . . . volatile alkali [ammonia] produced by means of nitrous acid and metals, Mr. Milner conceived the possibility of inverting the order of the process and of producing nitrous acid or nitrous air by the decomposition of volatile alkali. He knew of no experiments where this had been done, or any thing like it; yet . . . it seemed not unnatural to suppose, that by forcing volatile alkali through the red-hot calces [oxides] of some of the metals, nitrous acid or nitrous air might be produced; though in fact he neglected for near two years actually to make the trial. It was some time in the month of March, 1788, that the calx of manganese on account of its very great infusibility, and its yielding abundance of dephlogistigated air [oxygen], occurred as a very proper substance for the purpose. He immediately crammed a gun-barrel full of powdered manganese [dioxide]; and to one end of the tube he applied a small retort, containing the caustic volatile alkali. As soon as the manganese was heated red-hot, a lighted candle was placed under the retort, and the vapour of the boiling volatile alkali forced through the gun-barrel. Symptoms of nitrous fumes and nitrous air soon discovered themselves, and by a little perseverance he was enabled to collect considerable quantities of air [gas], which on trial proved highly nitrous. He afterwards frequently repeated this experiment, and always in some degree succeeded. Much depends on the kind of manganese employed, much on the heat of the furnace, and much on the patience of the operator; as these are varied, there will be great variations of the product.

The Reverend Mr. Milner tried other oxides and was hard put to it to explain why he had success with "calcined green vitriol" [ferrous sulfate] but not with "calcined alum." And while he admits that "it is better to acknowledge our ignorance than advance groundless speculations," he cannot resist speculating, in the turgid jargon of the day: ". . . it follows that the vitriolic acid has a stronger affinity to the inflammable principle than it has had to phlogisticated air; and the process with green vitriol and manganese is to be explained by the operation of a double affinity: the inflammable principle of the volatile alkali joins with the calx of iron, the basis of the vitriol, or with the manganese, and the phlogisticated air with the dephlogisticated air produced by the acid in the red heat. Those who chuse to reject the doctrine of phlogiston must make the necessary alterations in these expressions, but the reasoning will be much the same."

Fortunately for you and me, the reasoning will in any event be clearer and easier to follow!

Milner, as you see, produced nitric oxide by reaction between ammonia and a reducible oxide. The first account of its production by the reaction with which we are here concerned, the reaction of ammonia with oxygen in the presence of a platinum catalyst, was given by a French (Alsatian) chemist, Charles Frédéric Kuhlmann[4] (the founder of a chemical manufacturing concern bearing his name, still flourishing at the present time), in an oral report to the *Académie des Sciences* on December 24, 1838.

The first of twelve "novel reactions rendered possible by platinum sponge" was described by him as follows: "Ammonia mixed with air passing at a temperature of about 300°C over platinum sponge is decomposed, and the nitrogen which it contains is completely transformed into nitric acid, at the expense of the oxygen of the air."

He goes on to observe that one should not be greatly astonished that "in making use of a force which is not yet well known to us, which an illustrious chemist [Berzelius] has designated under the name of catalytic force, we could not easily foresee the results of our experiments."

Kuhlmann did not fail to recognize the economic and strategic importance of his observations. "By the facts contained in this note . . . I have made known the possibility of obtaining artificially and at will, nitric acid, and consequently, nitrates, without having recourse to the slow process of nitrification. If under present circumstances, the transformation of ammonia into nitric acid by means of platinum sponge and air does not offer any economy over present processes, time can come when such a transformation can become a profitable industry.

"One can say with assurance that knowledge of the facts I have stated is such as to set the country at rest completely about the difficulties or even impossibility of procuring saltpeter in sufficient quantity in case of a maritime war, and to cause to be totally

[4] *Comptes rendus*, **7**, 1107–1110 (1838); reported more fully in *Annalen der Chemie, Liebigs*, **29**, 272–279 (1839).

abandoned the former method of providing saltpeter for the needs of the State."

A good many years were to pass before the time came that Kuhlmann foresaw, when ammonia would become cheap enough to serve as a starting material to produce nitric acid. For it was nitric acid, not nitrates, that later was required for smokeless powder and high explosives. The capability of producing ammonia by the cyanamide process arose toward 1900, and ammonia liquor became available as the by-product in making coke for iron smelting.

Figure 3.1 Wilhelm Ostwald.
Courtesy Verlag-Chemie G.m.b.H.

It was at this period that Wilhelm Ostwald (Figure 3.1) of Leipzig, Germany, repeating Kuhlmann's experiments, came to the conclusion that the oxidation of ammonia was practical, and established the conditions for satisfactory operation. These were a platinum catalyst; an excess of oxygen; a high temperature, obtained by preheating the gases before reaching the catalyst; and a speed of passage over the catalyst just short of permitting ammonia to pass through unreacted. These conditions are clearly set out in his U.S. Patent 858,904, applied for June 26, 1902 (granted July 2, 1907).

Indeed, his preferred conditions, 10.6% NH_3 by volume, with the mixture preheated to 300°C, giving a calculated temperature after reaction of 1055°C, with a contact time of 0.01 sec, are well within the satisfactory operating range as established in subsequent industial practice. The process, therefore, to which this little book is devoted, the oxidation of ammonia with air over a platinum catalyst with the aim of producing nitric acid, has often been called the Ostwald Process. Ostwald received the Nobel Prize for chemistry in 1909.

A curious thing happened. The German Patent Office, not admitting any novelty in Ostwald's findings over those reported by Kuhlmann so long before, refused to grant him a patent, though he obtained patents in other countries, including the United States (after a delay of five years). Consequently, the Ostwald Process was practiced in Germany as a secret process. Beginning in 1908, it was operated at a colliery at Gerthe, in Westphalia, where by-product ammonia was converted to nitric acid. However, the plant was shut down in 1910, moved to Belgium, and began operation there in 1912.

The German military authorities were not even aware of the process at the outbreak of the war in the summer of 1914. Anticipating, as I have mentioned elsewhere, a quick and decisive victory, they had laid in a stockpile of Chilean nitrate of soda, and had made no provision for a long war. It was the eminent chemist Emil Fischer, having heard of the Gerthe plant by accident, brought the process to the attention of the military authorities in September 1914. They had begun to worry about a supply of nitric acid but had considered up to that time only the arc process, which would have consumed more electric power than Germany could possibly produce. At last convinced, the German authorities encouraged the Gerthe colliery to renew and expand its production, and the B.A.S.F. to undertake the development of an oxidation process to utilize the synthetic ammonia they were beginning to produce. Considering platinum too costly, the B.A.S.F. developed a base-metal catalyst, a mixture of iron and bismuth oxides. The government also encouraged the Frank-Caro group, who produced calcium cyanamide, in the development of a process, in which they did employ platinum wire gauzes, heated electrically (for reasons that will come to light shortly). By the end of the war, in 1918, the production capacity of the German plants had grown to several hundred tons of HNO_3 per day, derived of course mainly from synthetic ammonia, for lack of which the Germans undoubtedly would have been forced to capitulate much sooner.

Modifications to the original Ostwald process were introduced by another German, Professor Karl Kaiser of Heidelberg, who employed a gauze of platinum wire instead of Ostwald's crimped platinum foil. He also obtained a patent for the scheme of preheating the air for oxidation separately and mixing it with the ammonia just before entering the catalyst chamber, thus minimizing the contact of the ammonia with surfaces that would merely decompose it.

The Frank and Caro group, charged with developing an ammonia oxidation process starting with ammonia from calcium cyanamide, also realized that a higher temperature than attained by a cold gas mixture upon reaction was necessary to get good yields of nitric oxide. Perhaps because these chemists were conversant with electrical supply of heat the carbide furnaces and the cyanamide reactors, they chose to supply electricity to the platinum gauze catalyst to maintain its temperature in a favorable range. This may have been a fortunate choice for them, since ammonia from cyanamide contained small amounts of impurities, notably phosphine, PH_3, which acted as "poisons," as they are called, cutting down the activity of the catalyst. Keeping the temperature high by electric-resistance heating of the gauze, it appears, enabled the catalyst to recover its activity after occasional periods of "poisoning."

Ammonia Oxidation—American Installations

In any event, this modification, developed in Germany in 1915–1916, was adopted by the American firm manufacturing calcium cyanamide and demonstrated in an installation for supplying nitrogen oxides for a lead chamber sulfuric acid plant. This installation comprised six oxidation units, each with an electrically heated single-layer platinum gauze of about 2 sq ft area, found capable of handling about 10 lb/hr of ammonia.

After the United States entered the war in 1917, the Ordnance Department of the Army determined to have a plant built which would supply ammonium nitrate and avoid the necessity of transporting so much nitrate of soda from Chile, and they decided on a process already demonstrated as feasible. Accordingly, they authorized the construction at Muscle Shoals, Alabama, of a plant to produce 110,000 tons/year of ammonium nitrate, based on the cyanamide process. Never operated at full capacity before it was closed down upon the signing of the Armistice in November 1918, this plant, U.S. Nitrate Plant No. 2, was described in 1919 as "the largest synthetic nitrogen plant in the world."

The ammonia oxidation section, designed for the production of 250 tons/day of HNO_3, contained 696 of these electrically heated oxidizers, each expected to handle 10 lb/hr of ammonia.

Meanwhile, the Bureau of Mines had been asked to recommend a process for supplying nitrates in the apparently inevitable circumstance that the United States would be drawn into the war, and had come to the conclusion that the Haber direct ammonia synthesis and the Ostwald ammonia oxidation process offered the greatest promise. These processes had not been commercially established, however, in the United States, and the War Department was willing to authorize only a large-scale experimental plant, rated at 22,000 tons/year of ammonium nitrate, to be built at Muscle Shoals; this was designated as U.S. Nitrate Plant No. 1.

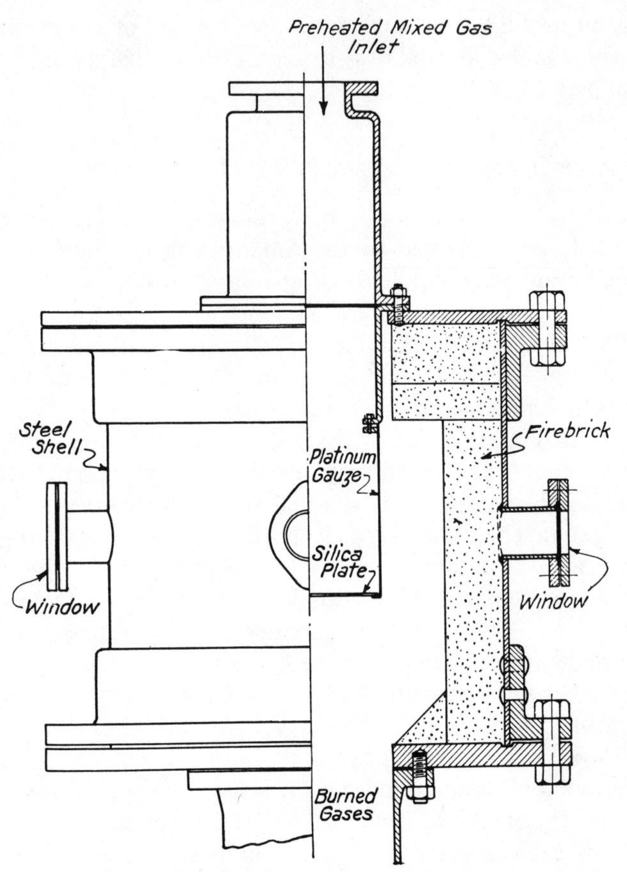

Figure 3.2 Circular gauze converter.

From H. A. Curtis, Editor, *Fixed Nitrogen*, American Chemical Society Monograph No. 59, Reinhold, New York, 1932, with permission.

The synthesis section never produced ammonia before the end of the war, but the oxidizers, operating on by-product ammonia liquor, were tried out and proved operable.

The oxiders were of a design proposed by Dr. C. L. Parsons, Chief Chemist of the Bureau of Mines (later the long-time Secretary of the American Chemical Society). The design goes back to Ostwald's original conception and relies upon heat exchange

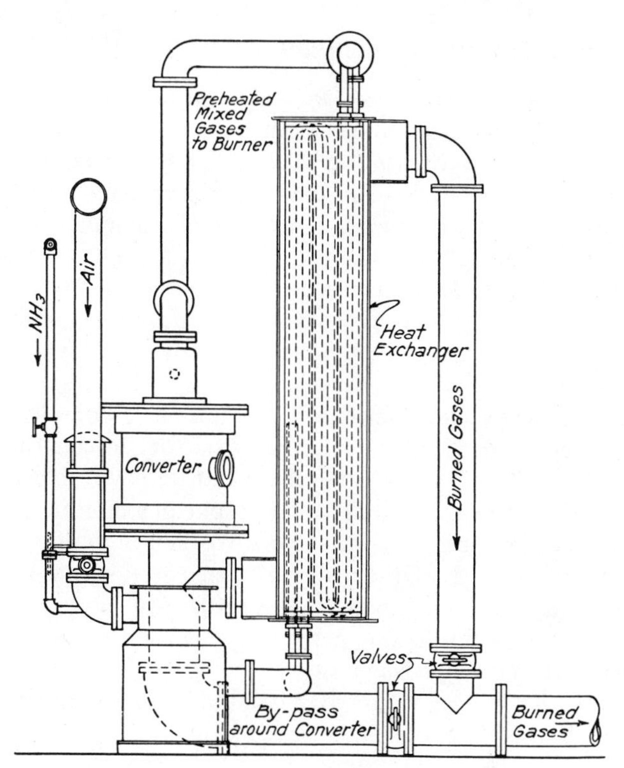

Figure 3.3 Commercial converter installation.

From Curtis, *op. cit.*, with permission.

with the products of combustion to bring the feed mixture to a suitable temperature level before coming in contact with the catalyst. A catalyst in multiple layers was found capable of handling much more ammonia than a single layer, through which some ammonia could pass unconverted. Dr. Parson's idea was that surrounding the catalyst, formed into a cylinder, with a refractory wall would stabilize its temperature and make it easier to control. Sketches of his converter, as supplied to H. A. Curtis[5] by Dr. Parsons, appear in Figures 3.2 and 3.3.

[5]H. A. Curtis, Editor, *Fixed Nitrogen*, American Chemical Society Monograph No. 59, Reinhold, New York, 1932.

Ammonia Oxidation—Chemistry

Such was the state of the technological development of the process of oxidizing ammonia to oxides of nitrogen at the close of World War I. Before taking up the technology of the production of nitric acid by recovery of the oxides, the chemistry of which was covered in Chapter Two, we will review at this point the chemistry of the catalytic oxidation of ammonia.

The oxidation of ammonia is unique. It proceeds rapidly, gives nearly 100% yield of nitric oxide under a wide range of operating conditions, and results in a product thermodynamically unstable at the temperature of its formation.

A process of such commercial importance has, naturally, been extensively studied, but no wholly satisfactory mechanism has yet resulted. The "force of catalysis," alluded to by Kuhlmann in 1838, is still imperfectly understood in spite of a multitude of researches and symposia on the subject. Workers in this field seem convinced, however, of the reality of chemisorption (the absorption of reactive species upon catalyst surfaces by loose or transitory compound formation), perhaps at "active sites" where unsatisfied chemical bonds appear.

From what you know already, it is unthinkable that the reaction would proceed as written in balanced form:

$$4NH_3 + 5O_2 = 4NO + 6H_2O. \tag{3.1}$$

Bimolecular reactions are what we expect; trimolecular collisions, even, are rare, and certainly a reaction among nine molecules is out of the question.

In addition to the main reactions, equations can be written for numerous other reactions that could be expected to take place simultaneously, or as seems more probable, consecutively. It is pretty obvious to a chemist that from any ammonia decomposed to the elements (see Equation 3.5) the hydrogen would react with any oxygen present, with the result shown in Equation 3.2. The lower oxide, N_2O, is found among the products (Equation 3.6) if the reaction is carried out at a low temperature, below about 500°C. At sufficiently high temperatures and sufficiently long contact times, nitric oxide would decompose, according to Equation 3.4. We recall from Chapter Two that only very small concentrations of this oxide exist in equilibrium with the elements at temperatures below about 2000°C. At the temperatures encountered here, however, 1000°C or less, the rate of decomposition even on a platinum catalyst turns out to be negligible at the short contact times that suffice for its formation in the oxidation of ammonia. Reaction 3.3 could occur, and perhaps does if contact with the catalyst is prolonged or, alternatively, is so short that some ammonia comes through unconverted.

$$4HN_3+3O_2 = 2N_2+6H_2O \tag{3.2}$$

$$4NH_3+6NO = 5N_2+6H_2O \tag{3.3}$$

$$2NO = O_2+N_2 \tag{3.4}$$

$$2NH_3 = N_2+3H_2 \tag{3.5}$$

$$2NH_3+2O_2 = N_2O+3H_2O \tag{3.6}$$

It can be seen that the reaction does not proceed according to any one of the equations shown so far. Any attempt to elucidate the mechanism of a reaction that goes as rapidly as this one must be somewhat speculative. In spite of the Reverend Mr. Milner's admonition about the futility of advancing groundless speculations, I will venture to bring together what seem to me the best-supported theories proposed by recent workers in this field. A sequence of steps such as the following can be drawn up:

$$NH_3+O \rightarrow NH_2OH \tag{3.7}$$

$$NH_2OH \rightarrow NH+H_2O \tag{3.8}$$

$$NH+O_2 \rightarrow HNO_2 \tag{3.9}$$

$$HNO_2 \rightarrow NO+OH \tag{3.10}$$

$$2OH \rightarrow H_2O+O \tag{3.11}$$

All these reactions, being bimolecular, could occur fast enough to account for the rapidity of the over-all reaction. The essential feature is the chemisorption of oxygen on the catalyst and the reaction with it of a molecule of ammonia, perhaps through the intermediate stage of hydroxylamine, NH_2OH, to produce a chemisorbed imide radical, NH (though this could be produced directly by loss of a water molecule from the complex $O{-}NH_3$). This imide radical then reacts with molecular oxygen to yield, eventually, nitric oxide and water.

Though this sounds fairly involved, I hope that it makes more sense to you than Milner's explanation of his observations in terms of the old phlogiston theory. Perhaps some day soon we will have a better understanding of catalysis than we do now.[6]

Actually, the limiting factor turns out to be not the rate of the chemical reaction sequence but the rate at which ammonia can be physically brought into contact with the catalyst, a problem that can be treated by well-known methods of chemical engineering analysis.

However all this may be, the observable facts remain much as stated by Ostwald. At a sufficiently high temperature and at a sufficiently short time of contact with the catalyst, with a suitable excess of oxygen, nearly quantitative conversion of ammonia to nitric oxide can be obtained. The upper limit on temperature is fixed by the loss of platinum from the catalyst. The temperature is controlled by the concentration of ammonia in the gas mixture and the extent of preheat obtained by heat exchange with the products of combustion, as discussed earlier.

The catalyst originally used was pure platinum, which gave good results. The presence of iridium was found to be harmful, but alloying with 10% of rhodium resulted in a distinctly superior catalyst, not only giving a little better conversion of the ammonia but reducing quite considerably the loss that the catalyst suffers on continued use. Palladium can be substituted for up to one-half of the rhodium, with almost equal performance, at a somewhat lower cost.

Representative results from an extensive investigation show conversions as high as 97.5%, with a four-layer cylindrical gauze catalyst of pure platinum, at atmospheric pressure, with 8.3% ammonia in air, preheated to give a gauze temperature of 930°C

[6]If you have an interest in learning more about catalysis, try to get hold of a copy of *Catalysis: Then and now*, by P. H. Emmett and P. Sabatier, translated from the French by E. E. Reid, Franklin Publishing Co., Englewood, N. J., 1965.

at a rate expressed as 100 lb NH_3 per troy ounce of catalyst per day (24 hours). With a 10% rhodium catalyst the conversion under the same conditions reached 99%.

The catalyst is conveniently provided in the form of a woven gauze of fine wires, generally of 0.003-inch diameter, 80 meshes to the linear inch (Figure 3.4). It must be clean, and it has to be

Before use After use

Figure 3.4 Platinum gauze catalyst (75X Magnification) before and after use.

Courtesy E.I. du Pont de Nemours & Co., Inc.

activated for a short period before it attains its full conversion efficiency. Originally consisting of shiny wires from the drawing operation, it takes on a blistered appearance after it has been in operation for a while. It is reported that electron microscope examination reveals a millionfold increase in the number of peaks on the surface after activation. Particles of platinum erode away, though some of them can be caught and recovered by a filter through which the gases can be put after partial cooling. An explanation has been offered of the changes occurring. During the oxidation of the ammonia, some hydrogen is formed, and this dissolves in the platinum and increases the mobility of the atoms in the crystal lattice and the ease of volatilization. Alternate oxidation and reduction break up the structure and increase the rate of erosion.

In any event, the catalyst does not last forever. The cost of catalyst replacement is not a major item, however. The loss is reported to be on the order of 0.005 troy ounce (1 troy ounce equals 31.1 grams) per ton of HNO_3 in a commercial installation. With platinum currently quoted at around $100 an ounce, this comes to only 50¢ per ton of acid, compared to a cost of about $25 for the ammonia required to produce a ton of HNO_3, at 95% over-all yield, and the current quotation of around $95.00 per ton. Fabrication of the gauze adds something of course to the cost of the catalyst, and the rhodium or rhodium-palladium alloy is somewhat more expensive than pure platinum.

Recovery of Nitrogen Oxides as Nitric Acid

We now turn our attention to the recovery of the oxides of nitrogen as nitric acid. The chemistry involved was treated in Chapter Two. It now remains to look at the means employed technically for the manufacture of nitric acid in the days before the development of stainless steel, that is, up to the end of World War I.

The primary factors determining the choice of equipment for the treatment of nitrogen oxides to produce nitric acid are chemical, on the one hand, and related to materials of construction, on the other.

The important chemical factors, as outlined in the preceding chapter, are the necessity of converting NO to NO_2 before it will react with water, the slow rate of this oxidation reaction, the incomplete extent of the reaction with water, and finally the favorable influence of low temperatures on the rate and equilibrium of these reactions.

The essential steps then are, in sequence: (1) Cooling the reaction products, after heat exchange with the gases going to the reactor, to a point which causes most of the water produced in the reaction to condense. (2) Allowing the gases, with admixture of some additional air to give an excess of oxygen over that required to convert the NO to HNO_3, enough time to cause as much of the NO as possible to oxidize to NO_2. (3) Then bringing the gases into contact in successive countercurrent stages with water, while allowing extended time of residence to reconvert to NO_2 the NO formed upon its reaction with water. Since as you read in the preceding chapter, this reaction goes more and more slowly as the concentration gets smaller, it never goes to completion, and some NO always escapes in the stack gases from the last stage of contact with water. A flowsheet is given in Figure 3.5.

The heat exchanger was never satisfactorily worked out during World War I. The electrically heated converter avoided this

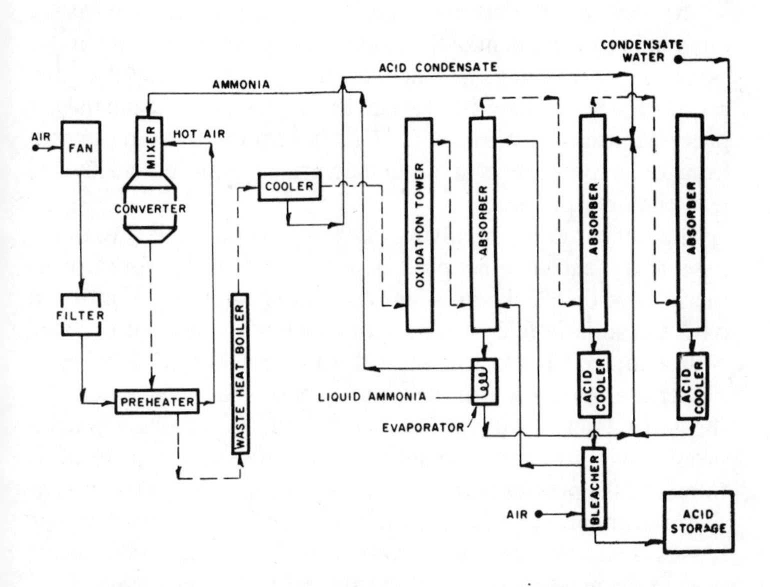

Figure 3.5 Atmospheric pressure ammonia oxidation process.

From V. Sauchelli, *Fertilizer Nitrogen: Its Chemistry and Technology*, American Chemical Society Monograph No. 161, p. 119, Reinhold, New York, 1964, with permission.

problem. Dr. Parsons, as you see in Figure 3.3, proposed to use aluminum tubes; but aluminum loses most of its strength at the temperatures required. A British firm, the United Alkali Co., used a converter coated with a cement made from barium sulfate and sodium silicate, to prevent contact of the ammonia-air mixture with the iron. It did not last very long.

Cooling down to a temperature approaching the dew point of the gases could be done in iron pipes, but from that point acid-proof stoneware was required, a poor transmitter of heat, and fragile, at that. High-silicon iron resists the acid, but it is sensitive to thermal shock.

The property of nitric acid, *aqua fortis*, of attacking all ordinary metals limited the choice of equipment that could be used to certain vitreous or stoneware compositions. However, this fact placed limitation on the pressure; the process could be carried out only at essentially atmospheric pressure.

The technology that developed in this period for recovering nitric acid by treatment of the oxides with water, under the limitations just mentioned, was derived from that employed in the retort process, as described earlier, and by the plants using the arc process for making nitric oxide. That in turn was derived perforce from older practice in the manufacture of sulfuric acid by the lead chamber process.

The hot tower for supplying "nitrose," as they called it, to the gases in the lead chamber process, and the cold tower, for recovering it from the stack gases, had to handle solutions of nitrogen oxides in sulfuric acid. Towers with a shell of lead were used, lined with acidproof brick, laid in acidproof cement, and filled with quartz lumps over which the acid flowed in making contact with the gases. Packings of variously contrived shapes, made of acid-proof stoneware, were later substituted for the quartz, in an effort to reduce the pressure drop of the gas flowing through the towers.

At the arc process plants the towers were built up out of granite blocks. Lead, somewhat resistant to sulfuric acid, could not be used in the presence of nitric acid alone. In American practice the towers were constructed of acidproof brick, filled with acidproof stoneware packings.

Details of the development of the technology employed in this now nearly obsolete process would be of small interest to you now. The process may be thought to have culminated in the plant built at Muscle Shoals, Alabama, U.S. Nitrate Plant No. 2, to which we already referred. A description of this installation will, I think, be illuminating, if only for giving a sense of the enormous dimensions required in a plant operating at atmospheric pressure. The plant was described in a report published January 1, 1919, by a consulting chemical engineer, A. M. Fairlie,[7] who is best known for his book on sulfuric acid manufacture.

In this plant designed for the production of 250 tons/day of HNO_3 at a concentration of 50% (for neutralization to ammonium nitrate), electrically heated catalyzers, 696 of them, as you were just reading, were supplied to convert the ammonia to nitric oxide. The gases leaving the converters passed through a cast-iron pipe into a flue made of concrete, lined with acidproof brick, to coolers and oxidizers. The first set of coolers was a battery of 24 steam boilers, ordinary steel, hot gases passing through the tubes, set to produce steam at 25-lb/sq in. gage pressure, keeping the steel well above the dew point of the gases. They were to cool the gases from about 600°C to only 200°C. The gases then passed through aluminum pipes to acid-resistant coolers, twelve in number, where "thimble tubes," as they are called by the chemical engineer, were suspended, closed at the bottom, each cooled by a stream of water supplied by an inner pipe reaching almost to the bottom, and overflowing at the top. The first two rows of these tubes were of chemical stoneware; the next three rows were made of high-silicon iron (Duriron). These were to cool the gases to 30°C; the condensed acid was drained off into the absorption tower system.

The gases then went to a set of twelve oxidation towers in parallel, each 15 feet square, divided up into four compartments. They passed finally to the countercurrent absorption towers, twelve sets in parallel, each set consisting of two acidproof brick structures, each 35 feet square and 60 feet high, again divided into four

[7] *Chemical and Metallurgical Engineering*, **20**, 8–17 (1919).

compartments, each packed with stoneware rings, giving eight countercurrent stages of contact. Aluminum fans moved the gases from one tower to the next and finally to the atmosphere. Acid was circulated over the towers by means of air lifts made of aluminum, set in wells 100 feet deep, like the ones you read about in Chapter One. The acid was to be cooled by water flowing around the air lifts.

The plant never operated for any length of time because the war ended. Perhaps it is just as well. Aluminum is not satisfactorily resistant to dilute nitric acid, and it is questionable how long the piping would have lasted. To provide against this contingency, they did provide 24 spares in addition to the 96 required for the absorption towers.

The important fact to grasp is that the gross volume of the absorption system came to the enormous total of 1,760,000 cu ft, not counting maybe 100,000 cu ft in the oxidation towers, for the design capacity of 250 tons/day of HNO_3 at 50% strength.

When you realize that most of this huge volume was required for the slow process of oxidation of re-formed nitric oxide back to the peroxide state, a process which can be speeded up in proportion to the square of the pressure, you see an incentive existed for the development of a process operating under increased pressure. That will be the subject of the next chapter.

CHAPTER FOUR

THE DU PONT PRESSURE PROCESS

During World War I, 1914–1918, the Du Pont Company[1] had heavy commitments for the supply of explosives to the United States Government and our Allies. In addition, Du Pont undertook to manufacture synthetic dyestuffs, the supply of which, from Germany, had been cut off by the naval blockade. An assured supply of nitric acid thus became of paramount importance. To this end Du Pont conducted its own operations in Chile to produce sodium nitrate, but ocean transportation was precarious, at least until the German submarine menace was overcome.

It was natural and inevitable, therefore, that Du Pont should give its attention to processes for nitrogen fixation. "Free," or elemental nitrogen, is freely and universally available, making up nearly four-fifths of the air. Oxygen makes up the remaining one-fifth, and these, with water, are the components of nitric acid. But the nitrogen of the air is inert, and the problem of "fixing" it, or bringing it into useful chemical combination, had for years challenged the inventive genius of chemists and engineers in many countries.

Their efforts, as recounted in the preceding chapter, had been directed along a number of diverse paths, among which there was one of special interest that involved the direct combination of nitrogen and oxygen at high temperatures, whereby a small pro-

[1]E. I. du Pont de Nemours & Co., Inc., Wilmington, Del.

portion of nitric oxide can be obtained. But the consumption of power in the electric arc used to give the high temperature was excessive for economic application in the United States.

The fixation of nitrogen by causing it to combine with hydrogen had been developed in Germany, as you were reading earlier, and appeared more economically attractive. Hydrogen could be obtained by reaction of water vapor (steam) with incandescent coke (or nowadays with natural gas in a catalytic "reformer"). The German process developed by Haber and Bosch had been modified by Georges Claude in France. It was this process, operating at about 15,000 lb/sq in., that the Du Pont Company chose to adopt. Arrangements were accordingly made with the French concern to furnish information that led to the construction of a plant at Belle, West Virginia.

Now that a process for nitrogen fixation that yielded ammonia had been adopted, it remained to select, or to develop, a process for converting this to nitric acid, preferably in high concentration, as required for nitration. The Du Pont chemists and engineers assigned to work on the problem[2] familiarized themselves with the technology of the ammonia oxidation process as then practiced at atmospheric pressure, and recommended the installation of a pilot unit at one of the explosives plants (Repauno Works, Gibbstown, New Jersey) to develop first-hand knowledge of the process. (See Chapter Five.)

Advantages of Operation at Elevated Pressure

Studying the published work available by that time on the chemistry of the nitrogen oxides, as reviewed in Chapter Two, the Du Pont staff became convinced that there was every advantage to be gained by operating the process under elevated pressure. In particular, the use of pressure would minimize the size of the plant and produce acid of higher concentration. Anticipating the commercialization of stainless alloys, known in the United States

[2]Notably, F. C. Zeisberg, F. C. Blake, and G. B. Taylor.

up to that time chiefly in cutlery, they determined to develop a process carried out under pressure. Tests showed that high-chromium iron, 16 to 17% Cr, was satisfactorily resistant to nitric acid.

It turns out that this was not the first time that the advantages of operating under pressure in the absorption of nitrogen oxides had been recognized. A British patent of 1919 refers to the use of pressure, but there is no record that the process described (involving refrigeration as well) was ever practiced.

Just to bring out again the kind of conclusions of industrial importance that can be deduced from laboratory physico-chemical data when available, consider the following. The controlling step in the production of nitric acid from the products of ammonia oxidation is the reaction between nitric oxide and oxygen. This reaction, you will recall, behaves as a classic third-order reaction, with a measurably slow rate constant. Now a third-order reaction is one where the reaction rate increases as the square of the pressure. The residence time for a given quantity of gas (by weight) in passing through a vessel of given volume increases in direct proportion to the pressure. It follows at once that the volume of the oxidation space to accomplish a given degree of oxidation of nitric oxide would be inversely proportional to the cube of the pressure. For a pressure of 8 atm, the volume required would be only $\frac{1}{512}$ of that necessary at atmospheric pressure—a tremendous reduction.

From the data on the equilibrium in the reaction of nitrogen oxides with water to form nitric acid, given in Chapter Two, you can easily estimate the strength of acid that could be produced from a given mixture of oxides as a function of the pressure. Consider, for example, a gas containing 5% total oxides by volume, and assume the same degree of conversion to nitrogen peroxide, say 80%, at a temperature of 25°C and at a total pressure of 1 atm; the equilibrium strength of acid would be about 55% by weight. At a pressure of 8 atm, the equilibrium strength would be nearly 65%, a very worthwhile increase if the acid must be further concentrated for its intended use in nitration processes.

Development of the Pressure Process

Experiments even on the smallest scale on a process such as this must be carried out on a continuous basis. The ammonia oxidizer set up for the first Du Pont trials, at the Experimental Station on the banks of the Brandywine Creek, consisted of a 3-in. square of platinum gauze in four layers, as employed in certain of the atmospheric pressure processes, heated by direct electrical resistance, held in nickel clamps, to maintain the temperature in spite of radiation losses. It was mounted in a 6-in. iron pipe covered with thermal insulation and provided with the necessary inlet and outlet connections. Air was supplied by a small compressor, and ammonia gas was taken off the vapor space of a cylinder of anhydrous ammonia, as supplied for commercial refrigeration. At 70°F ammonia has a vapor pressure of 114 lb/sq in. gage, so it was only necessary to set the cylinder in a tank of warm water to maintain the temperature at the low rate of draw-off used in these experiments, about 2 lb/hr.

The piping following the ammonia oxidizer was made of stainless alloys. Some tubing was available, but many fittings had to be improvised; such was the stage of development at that time. A 3-in. pipe provided the space for the oxidation of the nitric oxide. The absorption system consisted of superposed bubblers made of pieces of 3-in. high-chrome iron pipe with the ends welded, with perforated $\frac{3}{8}$-in. pipe for the introduction of gas beneath the surface of the liquid.

The very first experiments carried out with this equipment, at pressures of 80 to 100 lb/sq in. gage, confirmed, even more strikingly than anticipated, the effect of increased pressure on the absorption end of the process, while showing that acceptable conversions of ammonia to nitric oxide could still be obtained. The condensed water resulting from the conversion reaction was found to have reacted with the higher oxides of nitrogen in as short a time as could be technically provided, producing a rather strong acid that could be added to the absorption system at an appropriate point. In runs lasting a few hours at a time, conversions of ammonia to nitric oxide up to 85% of theoretical, and acid strengths to 65% HNO_3 by weight, were obtained.

These results were distinctly encouraging, but a process is not considered as demonstrated commercially until weighed quantities of product have been secured from weighed amounts of raw material, in continuous runs.

The next logical step in the development of the process, therefore, was a demonstration semiworks unit, as it is called, capable of continuous operation and over-all determination of acid yield. The pressure range chosen was up to about 100 lb/sq in. This was not arrived at by any extensive optimization procedure such as might now be carried out with the aid of electronic computers. It was a pressure for which standard-size pipe and fittings were usable; at which ammonia could be handled conveniently; and at which the calculated dimensions of the absorption system came down to about what would be needed for effective contacting of liquid and gas in any event.

A diagrammatic sketch of the semiworks unit is shown in Figure 4.1, and will serve as what chemical engineers call a "flowsheet" of the process.

Liquid anhydrous ammonia, taken off the liquid space of a cylinder, was vaporized by steam and somewhat superheated before being metered into the system. Air from a compressor was metered, then heated in an exchanger with the products of the ammonia oxidation reaction, and filtered through an asbestos fiber pad before being mixed with the ammonia and fed to the converter. Filtering is necessary to protect the gauze catalyst from contamination. Preheating is necessary to maintain the reaction at a favorable temperature, namely, in the range of about 900 to 1000°C, as pointed out in the previous chapter. Some excess of air over the stoichiometric proportion, 14.3% NH_3 by volume, is necessary both to obtain good conversion and to avoid operating close to the explosive range, for which the lower limit is about 16% by volume. A concentration of 9 to 10% gives favorable conversions. As the temperature rise on reaction for a 10% mixture is only about 700°C, the equivalent of about 200°C preheat must be provided. Exposure to metals, particularly iron, causes decomposition of ammonia at elevated temperatures, so it is desirable to preheat the air separately and mix it with the ammonia just before

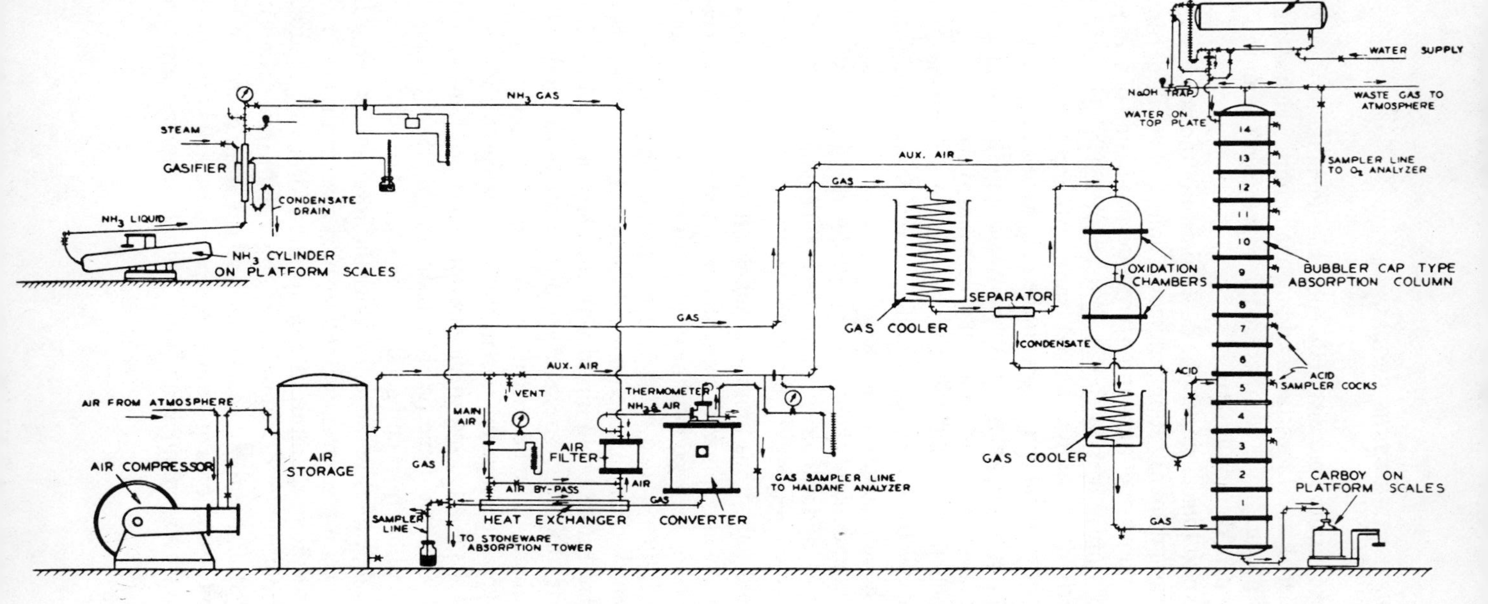

Figure 4.1 Diagrammatic sketch of semiworks unit.

From *Industrial and Engineering Chemistry*, *23*, 860, 1931, with permission of the American Chemical Society, copyright owners.

it enters the converter. The mixer was made of nickel, a much less harmfully catalytic substance than iron. Aluminum is even less active catalytically but is not strong enough at the temperatures involved. Stainless alloys are usable.

The converter was made of a piece of 12-in. iron pipe, flanged, with a refractory lining. The catalyst was a 3-in. cylinder of four-layer platinum gauze, 6 inches deep, fastened to the end of a piece of nickel pipe and closed off with a fused quartz disk, similar to the one shown in Figure 3.2.

From the converter the gases passed through the heat exchanger and then downward through a coil submerged in a tank of water. Here acid, 35 to 55% HNO_3, was condensed and trapped into an appropriate plate of the absorption column. The gas, plus auxiliary air sufficient to supply a slight excess of oxygen over that necessary to convert the oxides to HNO_3, passed through empty vessels serving as an oxidation space and thence to the base of the column.

The column was made of chromium-iron castings, 15 inches in diameter, with one 6-in. bubbler cap per plate. A plate efficiency, obtained by calculation methods based on equilibrium data given in Chapter Two, of around 50% appeared to be obtainable. It is unlikely, in view of the complexity of the mechanism of the absorption reaction, that a constant value prevails over the whole column. It has been possible, however, to approximate the performance of plant-scale absorption systems by using this procedure.

At the semiworks stage of development the pertinent variables were fully explored. The capacity was varied from about 6 to 12 lb/hr of ammonia, generally 10 lb/hr, corresponding to about 750 lb/day of HNO_3. The pressure range studied was from 50 to 100 lb/sq in. gage. The conversion efficiency was found to fall off somewhat with increasing pressure, but the absorption efficiency increased as expected. Over-all recovery of 85% of theoretical was obtained, fully justifying the economics of the process. (Yields over 90% are obtained regularly on the commercial scale.)

As just mentioned, an optimum operating pressure might have

been arrived at on the basis of extended trials and computations, but a converter pressure of around 100 lb/sq in. was selected for the design of larger plants. This was judged to be the highest justified by the effect of pressure on the size of the absorption system, while still requiring only standard compressors and vessels of standard specifications.

Later, more extended analysis using computer techniques still confirms this choice of operating pressure as near the economic optimum. Calculations summarized in Table 4.1 give an indication of the effect of moderate changes in operating conditions. Increase in plate spacing reduces the number of plates required but not the over-all column volume. Reducing the temperature reduces the number of plates, but only slightly; this leads to the

TABLE 4.1
Number of Theoretical Stages Required for 98.2% Recovery of Nitrogen Oxides

Column 5 feet in diameter. Gas feed rate 1000 lb-moles/hr, containing 6.0% oxygen, 0.34% nitric oxide, 6.82% nitrogen peroxide. Acid produced 60.0% HNO_3 by weight, containing 5742 lb/hr HNO_3, of which 1316 lb/hr is formed in the cooler-condenser.

Effect of Plate Spacing

Pressure, 7 atm; Temperature, 30°C

Plate Spacing, ft	No. of Plates
1	20
2	15
3	12

Effect of Temperature

Pressure, 7 atm; Plate Spacing, 1 ft

Temperature, °C	No. of Plates
20	18
30	20
40	28

Effect of Pressure

Temperature, 30°C; Plate Spacing, 1 ft

Total Pressure, atm	No. of Plates
5	57
7	20
9	12

conclusion that a large amount of heat transfer surface to take out heat by cooling water is desirable but that refrigeration would not be economic. Reducing the operating pressure increases the column requirements, but increasing the pressure looks as if it would very soon reach the point of diminishing returns.

On the basis of the performance of the semiworks unit at the Experimental Station, a pilot plant with fifteen times its nominal capacity was built at the Repauno Works. The flowsheet was essentially the same. The cooler-condenser was of the return-bend type with water trickling over the bank of tubes. Enough progress had been achieved in the fabrication of stainless alloys to permit the absorption column to be made up with a riveted shell of high-chrome iron, and conventional bubble-cap plates of the same material. The column of twenty plates 45 inches in diameter and 35 feet high was provided with two cooling sections. In view of the unfavorable effect of high temperature on both the rate and equilibrium factors, it is essential on the large scale to cool both the gas and the liquid to remove as much as possible of the considerable heat of reaction, as pointed out before.

This plant made 4 tons/day of HNO_3 at 60% strength during the hot summer months and 5 tons/day at 63% during the winter, due to the favorable effect of low water temperatures. The overall yield proved to be close to 90% of theoretical. The loss out the stack was seldom as high as 2%, normally 1.5%. This could be reduced fractionally by more plates, with oxidation space between, as might be justified by balancing the fixed charges on the investment on additional sections against the value of the equivalent ammonia lost. It would hardly justify alkali treatment. (See Chapter Five.)

After this plant had served its original purpose as a demonstration unit, it was moved to another location, where it served the production needs for a number of years.

The lower indicated investment cost and the stronger acid resulting from pressure operation more than counterbalanced the cost of power for compression and the somewhat lower conversion efficiency. There resulted a net advantage over the atmospheric pressure process, on which operating experience had been obtained concurrently with this development. Accordingly, the pressure process was adopted for subsequent installations required to supply the Du Pont Company's needs for nitric acid.

Eight units, each with a nominal design capacity of 10 tons/day of HNO_3, were installed at Repauno. Initially, these units each had two cylindrical-gauze converters, the same as employed in the 5-ton/day pilot plant. Subsequently it was found that a single 2-foot diameter converter with a flat gauze in multiple layers, supported on a Nichrome grid, was capable of handling enough NH_3 for 25 tons/day HNO_3 at 96% efficiency. A catalyst of 90% platinum–10% rhodium was developed which gives these conversions with much less loss of catalyst than does pure platinum. More recently it was demonstrated that some of the rhodium could be replaced with lower-cost palladium with no decrease in efficiency.

The columns in these units were 5 feet 3 inches in diameter and 40 feet high. The problem in the absorption system, as mentioned earlier, is as much one of heat transfer as of contacting efficiency. Increased heat transfer surface in the form of multiple cooling coils laid in on the plates of the absorption columns sufficed to bring the absorption capacity of each of these units up to the 25 tons/day HNO_3 capability of the converter.

These units have given an over-all yield of 93% of theoretical, at an acid strength of 61%, with not more than 2% loss in the stack gases.

These steps in the development of the pressure process, from experiments on the scale of 150 lb/day to multiple units with a capacity of 50,000 lb/day each, have been recounted in some detail as illustrating the primary work of the chemical engineer: taking a process as conceived in the laboratory and bringing it to full-scale economical production.

Subsequent History of the Pressure Process

Subsequent development of the pressure process has involved mechanical engineering and metallurgy as much as chemical engineering. Many features of the process were patented by Du

Pont, and licenses issued under these patents were granted to a large number of companies.

An installation described in 1952 produces 50 tons/day HNO_3 at 60% strength. The flat-gauze converter uses 105 troy ounces of platinum-rhodium gauze catalyst. The absorber is 6 feet in diameter, 40 feet high. Intensive cooling is provided; three layers of cooling coils on each of the plates in the lower section, two layers in the intermediate section, and one in the top section.

The flowsheet of the process resembles that of the original semiworks unit in most respects. The important addition is a power-recovery compressor that furnishes part of the air supply —about 40% in this installation. This depends for its success on reheating of the exhaust gases to about 250°C, by further heat exchange with the gases from the ammonia oxidizer. This reheat serves both to increase the available energy in the exhaust gases when they are expanded to atmospheric pressure and to prevent condensation and corrosive attack on the blades of the expansion turbine (or on the reciprocating expansion engine, if that type is used).

Much larger units have been built. One described in 1956, at that time the largest in the United States, is rated at 220 tons/day HNO_3, at 58% strength. Among design features incorporated is an activated alumina filter for vaporized ammonia to prevent contamination of the gauze. The ammonia is oxidized in 10% by volume concentration in air, at 110 lb/sq in. gage pressure, preheated to 280°C, giving a gauze temperature of about 935°C. The gases then pass through a stainless heat exchanger containing two U-tube bundles, one at each end. The hot gases give up heat first to the exhaust gases coming from the absorption column and then to the air passing from the compressor to the gas mixer and converter. The cooler-condenser is arranged for upward flow of the gases countercurrent to water trickling over the coils.

The column is 10 feet in diameter, 50 feet high. The gases leave the column at about 80 lb/sq in. gage pressure and are reheated to 500°C or higher before reaching the expansion turbines. The turbines, of which there are two in parallel, run at 6500 rev/min, with eleven stages. They are direct-connected, along with a steam turbine that supplies about one-third of the power, to a two-case centrifugal compressor. In one compressor five stages compress the air to 40 lb/sq in., and in the second four stages compress it further to 120-lb/sq in. gage, each compressor handling 12,000 std cu ft/min.

The net power requirements are stated to be 160 kw hr/ton HNO_3, and the catalyst loss is only 0.005 troy ounce/ton.

More recent units, rated at 250 tons/day, incorporate a waste-heat boiler that supplies the steam for the turbine furnishing power to the compressor not supplied by the turbine expander. The flowsheet still contains the same elements as before, with the addition of a platinum recovery filter, and of the waste-heat boiler, with superheater and feedwater heater, all drawing heat from the ammonia oxidizer gases. The steam so produced, fed to the steam turbine, suffices to supplement the energy available, some 3100 kw, in the reheated waste gases passing to the expansion turbine and makes the plant, once started, self-sustaining as regards power for air compression.

Some details may be of interest on these large units installed by one of several firms prepared to design and build plants embodying the basic concepts of the Du Pont pressure process. The converter, Figure 4.2, employs 36 layers of catalyst gauze, in hexagons 36 inches across, weighing approximately 500 ounces (Figure 4.3). The expander and compressor, Figure 4.4, are specially selected for this service. The compressor, rated 25,000 cu ft/min at 120 lb/sq in. gage, uses, as before, nine stages of compression. Illustrations showing heat exchangers, Figure 4.5, the absorption column, Figure 4.6, with 35 bubble-cap plates, arranged as shown in Figure 4.7, and the tail gas preheater, Figure 4.8 (not all at the same location) suffice to indicate the magnitude and compactness of the unit. The plant gives a yield of 92.5 % of theoretical at an acid strength of 57 to 58 % HNO_3.

The pressure process, pioneered by Du Pont, with the modifications and improvements just described, has come to be the prevailing process in the United States, accounting for 61 out of 68 plants (1964 figures), or 90 %, and in total capacity 5,775,000 tons/year out of 6,400,000 tons/year HNO_3. There are also numerous installations in other countries.

Figure 4.2 Converter for 280-ton/day nitric acid plant.
Courtesy The Chemical and Industrial Corp., Cincinnati, Ohio.

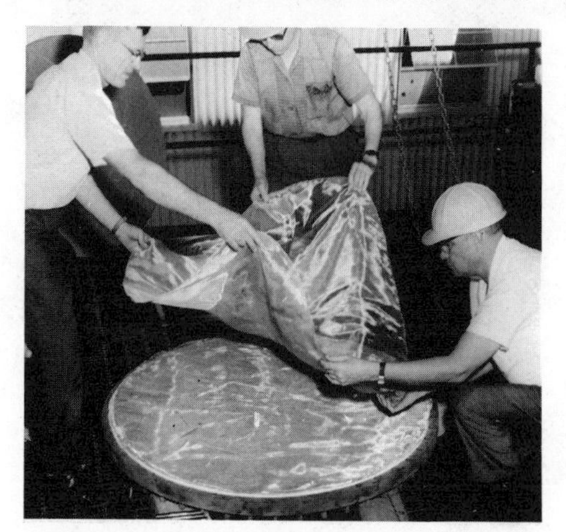

Figure 4.3 Installing platinum alloy gauze in converter.
Courtesy International Nickel Co., Inc., New York.

Figure 4.4 Expander and compressor for 280-ton/day nitric acid plant.
Courtesy The Chemical and Industrial Corp., Cincinnati, Ohio.

Figure 4.5 Heat exchangers for 250-ton/day nitric acid plant.
Courtesy The Chemical and Industrial Corp., Cincinnati, Ohio.

Figure 4.6 Absorber for 250-ton/day nitric acid plant.
Courtesy The Chemical and Industrial Corp., Cincinnati, Ohio.

Figure 4.7 Interior construction of nitric acid absorber.
Courtesy Fritz W. Glitsch & Sons, Inc., Dallas, Texas.

Figure 4.8 Tail gas preheater for 250-ton/day nitric acid plant.
Courtesy The Chemical and Industrial Corp., Cincinnati, Ohio.

While developed and established, the Du Pont pressure process cannot be said to have been fully perfected. A paper[3] published in 1964 from the Tennessee Valley Authority shows how, in accordance with physicochemical principles set forth in Chapter Two, the acid strength can be increased by providing a longer time at a lower temperature for increasing the degree of oxidation of nitric oxide to nitrogen peroxide before the gases enter the absorption system.

[3] *Chemical Engineering Progress*, **60**, No. 7, 77–84 (1964).

Further modifications and refinements are embodied in a plant in Mississippi,[4] started up in 1963, rated at 280 tons/day, claiming an over-all yield of 96% of theoretical, producing acid of 57% concentration. A plant in Tennessee was to have a capacity of 500 tons/day.[5] Beyond this, the unit being constructed at Louisiana, Missouri, by Hercules, Inc., is the largest on record: 800 tons/day of HNO_3.[6]

I think I ought to include here a few words on safety and related topics.

Anhydrous ammonia, the oxides of nitrogen, and nitric acid are all dangerous chemicals, but through the exercise of reasonable precautions the safety record of the plants producing them has been good. The oxidation mixture is operated fairly close to the lower explosive limit, but the potential explosive hazard is not great, and there have been instances where, at atmospheric pressure at least, the mixture has gone through the explosive range without anything untoward happening.

The absorption towers and piping have been remarkably free from corrosion in spite of the property of nitric acid to dissolve almost everything metallic. The passivating effect, interpreted as continuous renewal of a protective oxide film, on the surface of properly constituted and properly heat-treated chromium alloys, has been reliable in preventing attack under most circumstances.

One precaution must be observed. Chlorides must not be allowed to enter the column in the water used for absorption or in the air supply to the converter. Hydrochloric acid would be produced, and it would react with nitrogen oxides in the base of the absorption column to produce nitrosyl chloride NOCl, which is volatile and tries to work its way back upward, but it is hydrolyzed by water in the upper sections. The hydrochloric acid of course has too low a vapor pressure to be carried out of water solution with the exit gases. Consequently it builds up in intermediate sections and can become dangerously corrosive because of its ability to penetrate the protective film. The use of distilled

[4] *Chemical Engineering*, 38–40 (March 30, 1964).
[5] *Ibid.*, 116–117 (May 23, 1966).
[6] *Evening Journal*, Wilmington, Del. (Sept. 29, 1965).

water (steam condensate) as feed to the column and periodic checks of chloride concentration at several points in the column suffice to avoid this difficulty.

The fumes from the exit stack of a large plant are conspicuous on account of their red-brown color, though perhaps no more dangerous or even irritating than those from some other common chemical or metallurgical operations. Fume-abatement procedures have been developed. One consists in adding fuel gas to the preheated effluent and passing the mixture through a bed of crimped Nichrome ribbon on which a noble metal mixture has been deposited. The reaction produces only elementary nitrogen, carbon dioxide, and water. The heat of reaction is recovered in an additional waste-heat boiler.

For a full discussion of fume abatement, see Public Health Service Publication 999-AP-27, "Atmospheric Emission from Nitric Acid Manufacturing Processes."

I might cite one figure to point up the over-all economy of the Du Pont pressure process as compared with the atmospheric pressure process as developed up to 1925. I had occasion during the World War II to estimate the saving in installed cost of the nitric acid plants built in the United States for the production of (conventional) military explosives over the probable cost of plants built on the basis of World War I technology, that is, the atmospheric pressure process, except for the use of stainless steel construction. While it is unthinkable that there would not have been some advance in technology in the interval, nevertheless the basis for such a comparison was available. The resulting figure for the saving to the U.S. Government attributable to the development of the pressure process came out to be, roundly, $250 million.

Other processes for the manufacture of nitric acid will be described in the next chapter.

CHAPTER FIVE

ALTERNATIVE PROCESSES

You read in the preceding chapter that the Du Pont pressure process is the prevailing process for nitric acid manufacture in the United States. There are others that compete with it, however, and I will proceed now to discuss some of these alternative processes.

Atmospheric Pressure Plants

The atmospheric pressure process as originally proposed by Ostwald is not quite obsolete. Early in the study of ammonia oxidation processes, the Du Pont Company, just to gain experience with the process, built at the Repauno Works a pilot unit using a Parsons converter and square, acidproof brick towers similar to the huge ones at Muscle Shoals. It was operated by the plant personnel under the supervision of Eastern Laboratory for about a year, and performed well, as anticipated.

Aside from affording experience with the process, the unit yielded one important result. Ahead of the brick and stoneware absorption towers, they installed a riveted high-chrome iron vessel, used as an absorber, the first produced in the United States. The highly satisfactory experience with this vessel resulted in the prompt decision to use stainless alloys exclusively in the construction of a plant of 25 tons/day capacity, being designed to keep pace with the production of synthetic ammonia, even though it was to be operated at atmospheric pressure, pending demonstration of the pressure process. This plant came into operation in 1927. The availability of stainless alloys for pumps, piping, and heat transfer equipment was an enormous advantage in safety and continuity of operation over the older stoneware systems. The plant fully justified itself, though subsequent installations have all with one exception been of the 100-lb/sq in. system.

The plant had five Parsons type converters, stainless heat exchangers, and gas coolers. Several empty oxidation towers preceded the absorption towers, which were eight in number, 10 feet in diameter, 50 feet high, packed with acidproof stoneware rings. Even with such a large volume, approximately 40,000 cu ft, it was not expected that an economic recovery of nitrogen oxides would be obtained by treatment with water alone. The last two towers were intended for final "scrubbing" of the gases with alkali (sodium carbonate) to produce sodium nitrite. You recall from Chapter Two that ($NO + NO_2$) reacts with alkalies to yield nitrites. By careful control of the alkalinity of the scrubbing solution, never allowing it to become acid, and making sure that the gases entering had some excess of nitric oxide over NO_2, quite pure nitrite was produced.

The plant was operated after this fashion for some time, but with nitric acid being supplied by other, larger units, and the market for nitrite increasing, it was converted principally to the production of nitrite, and in this role it continues in operation.

It is in such circumstances that the atmospheric pressure process remains competitive. Sodium nitrate can also be produced when and where the market for it exists, by treating the alkaline scrubbing solutions with nitric acid made in the first few towers of the system, and returning the oxides produced by the decomposition of the resulting nitrous acid back into the absorption system.

In England, Imperial Chemical Industries, Ltd. (I.C.I.), who early installed a Du Pont unit, has several large units in operation, each rated at about 100 tons/day, making acid of 50% strength, with alkali scrubbing to produce nitrite, working just above atmospheric pressure, 3 lb/sq in. gage.

Intermediate Pressures

Moderate or intermediate pressures have been selected for some installations, generally those where the product is used to make ammonium nitrate, and a high HNO_3 concentration is not needed. One such design, developed by the Dutch State Mines (Staatsmijnen), is represented in the United States by a plant recently built in Georgia rated at 470 tons/day at 57% strength, operating at 3 atm pressure. The absorption system consists of three towers, each containing several packed sections, over which acid is circulated separately, with cooling. In the first tower the gases flow downward, and it is called the oxidation tower. High absorption efficiency is claimed, as well as higher yield of nitric oxide in the ammonia oxidation step, and a lower catalyst loss than in the 8-atm (Du Pont) process.

The published tabulation of cost factors, Table 5.1, in comparison with those attributed to the conventional high-pressure process is interpreted as showing a distinct competitive advantage

TABLE 5.1
Cost Comparison

	Medium-Pressure Plant	High-Pressure Plant
Capital costs, $/(ton)(day)	5.20	5.10
Catalyst consumption, grams/ton	0.099	0.190
Power, kw hr/ton HNO_3	15	10
Export steam, lb/ton HNO_3	800	250
Cooling water, gal/ton HNO_3	30,000	35,000
Ammonia yield, %	95	93

for the medium-pressure process. (But you might compare the yield claimed with that claimed for the high-pressure plant in Mississippi, mentioned a few pages back. Variations in yield are probably the most significant cost item.)

For a plant of their own design erected in 1957, I.C.I. chose operation at 3 atm to produce acid of 60% strength. It incorporates a number of refinements, including even the utilization of some of the heat given off in the oxidation of NO to NO_2 for the reheating of the exhaust gases before expansion in a power recovery turbine. Some of the heat given out in the cooler-condenser is used to vaporize the ammonia feed. A rather high concentration of ammonia is used in the feed mixture, 11% by volume, so that only a moderate preheat is needed. The ammonia alone needs to be preheated in an aluminum alloy exchanger and mixed with air merely warmed by compression and filtered. The absorption column is provided with "sealed" sieve (perforated) plates, thought to be more efficient than bubbler-cap plates used in the conventional (Du Pont) design. Performance figures have not been reported.

At the other end of the scale in daily output is an installation in New York State. It is rated at only 1 ton/day of HNO_3, at 60% strength, operating at 50 lb/sq in. gage, on aqua ammonia. It cost about $50,000. It was expected to pay for itself, however, in about one year, in savings over the cost of buying acid in carboys (as glass bottles of large size are called).

Combination Processes

Several plants have been built in Europe to a design worked out by the Dutch State Mines which attempts to obtain the advantage of oxidation at high conversion efficiency at atmospheric pressure with the high acid strength and high absorption efficiency obtained and reduced absorber volume required under higher pressure, by the device of compressing the gases between the ammonia oxidation converter and the absorption system. I.C.I. Ltd. chose an installation of this type for one of their plants, described in 1961, a plant where high-strength acid is needed for the manufacture of explosives.[1] The plant capacity is not stated, but the flat gauze in each of three converters is 9 feet $1\frac{1}{4}$ inches in diameter. Cooling is done by means of a waste-heat steam boiler (ordinary steel) followed by water cooling to below the dew point in stainless alloy equipment. The condensate is weak and is fed to

[1] *Industrial Chemist*, **37**, 159–166 (1961).

the absorber system. A rather high percentage of ammonia in the gas feed to the oxidizer, 11.5%, reduces the preheat needed, and preheating is done with steam at 300°C from the waste-heat boiler.

The cooled gas stream plus necessary secondary air is compressed in a high-chrome iron centrifugal compressor to 42 lb/sq in. gage (about 4 atm). The heat of compression is utilized to warm the exhaust gases before they reach the expansion turbine that helps drive the compressor, principally driven by a steam turbine. After further cooling, the gases go to an oxidation tower packed with rings, with cooled acid circulated over each of four stages, then to a succession of five packed absorption towers, all 8 feet in diameter, 62 feet high.

Satisfactory performance is reported, good conversion efficiency, low stack losses, and acid of 60% strength. I.C.I. has chosen the same process for a larger plant, rated at 570 tons/day, at another location.

What is known as the Montecatini process follows the same general pattern, but for the absorption system (operating at around 40 lb/sq in. gauge) it employs a series of horizontal tanks. Said to be well established in Europe, it is in use in only one installation in the United States,[2] the one at the Geneva works of the United States Steel Corporation, at Provo, Utah, rated at 160 tons/day of HNO_3. The converter combines the catalytic gauze in the same unit with a waste-heat boiler and superheater. The oxidizer gas then passes through a heat exchanger to reheat the exhaust gases going to the turbine expander driving the air compressor, through an economizer, giving up heat to the waste-heat boiler feed water, and through a shell-and-tube cooler-condenser. The condensate is dilute enough to be fed to the last stage of the absorption train.

The absorption system is unique. It consists of twenty horizontal stainless steel vessels 6 feet in diameter, 41 feet long. A pool of acid is maintained in the bottom of each, fed forward from one to the next countercurrent to the gas stream. The gases bubble

[2] *Chemical Engineering*, **65**, No. 9, 56, 58 (1958).

through this pool from a perforated longitudinal channel. The acid is sprayed against the walls of the tanks, which are cooled by a flow of water over the exterior. The principle is the same as in the Du Pont process, the space above the acid serving to reoxidize the nitric oxide produced in each absorption stage before it passes to the next. The volume of the absorption system calculates as 23,200 cubic feet for the capacity stated, as compared with 3000 cubic feet for a Du Pont plant of the same rating.

The Montecatini plant operates with a somewhat richer gas mixture as compared with the Du Pont process and does not employ preheating of the mixture. As it operates at about 815°C, lower platinum losses are claimed, only 0.0025 troy ounce/ton HNO_3.

No external power is said to be required if the steam produced is employed to help drive the turbocompressors (not done in the installation described). Over-all yield is claimed to be 94 to 95%, with 97 to 98% yield in the converter.

Another variation on the combination process is that employed by Etablissements Kuhlmann in France. Ammonia is oxidized at atmospheric pressure and the products are compressed to about 40 lb/sq in. gage. Four absorbers in parallel are used for a capacity of 200 tons/day, each with sixteen plates 15 feet diameter and 70 feet high. The unusual feature of the design is the division of the plates into sectors, with the overflow pipes staggered around the column from plate to plate, and a stagnant pool on each plate which is said to permit reoxidation of the gases. Cooling by refrigeration is employed, and acid strengths to 70% and 98% yield are claimed.

A number of installations have been made based on this process, mostly in Europe, and one in Japan, some rated as high as 380 tons/day HNO_3 at 70% strength.

One further alternative scheme has been adopted by a French concern known as Grande Paroisse. They carry out the ammonia oxidation at 35 lb/sq in. gage, followed by compression of the nitrous gases, after cooling to 78 or to 93 lb/sq in. gage for absorption. They thus achieve some compactness in the ammonia oxidation equipment and reduce the load on the nitrous gas compressor. They strive for heat economy at every point, even using as a boiler feed-water economizer the interstage cooler between the two stages of the air compressor. The heat produced in the second stage is used as preheat for the converter. Vaporizing ammonia at about the freezing point of water, they obtain some refrigeration to use in the absorption system. Providing ample, cooled oxidation space for the gases before entering the plate-type absorption tower, they are able to get a high-strength acid, claiming 70%. An interesting feature is the coupling of five units on one rotating shaft: two stages of air compression, one of nitrous gas compression, the expansion turbine, and the steam turbine. Complete self-sufficiency of power is claimed, even leaving some steam for use elsewhere in the plant.

So you see that opportunity still remains for development in the ammonia oxidation process. A choice of processes remains open, with selection influenced by the strength of acid required (whether it is to be used for ammonium nitrate, for instance, or whether it must be concentrated for use in nitration reactions), the cost of ammonia at the location chosen, the cost of power, as well as by the plant investment. As for this factor alone, a published high-spot comparison[3] concludes that there is only a ±11% difference in the installed plant investment among the processes so far described, running in about the following relative proportion:

Atmospheric pressure	1.0
Pressure process	0.89
Combined (atmospheric oxidation, pressure absorption)	1.11

Direct Production of High-Strength Acid

The Bamag-Meguin process is based on a different principle from the others considered here. Ammonia is oxidized with oxygen in only slight excess at atmospheric pressure; steam is added to take up the heat of reaction and keep below the explosive range. The excess steam and the water produced in the reaction are condensed and discarded, while the extent of oxidation of the

[3] *Chemical Engineering*, **63**, No. 5, 170 (1956).

nitric oxide is regulated to limit formation of NO_2 at this point.

The gases are then further cooled and oxidized, and the remainder of the water is removed as nitric acid. The gas now consists mainly of nitrogen peroxide (boiling point 21.15°C), which is condensed out by refrigeration. An absorption column recovers oxides from the residual inert gas.

The liquid nitrogen peroxide is mixed with the nitric acid produced at these other points, made up to contain enough water to react with the N_2O_4 according to the equation

$$2N_2O_4+O_2+H_2O \rightarrow 4HNO_3,$$

and is pumped to a pressure vessel known as an autoclave, which in recent installations is operated in continuous fashion. The pressure is maintained at 52 atm, 750 lb/sq in. gage, and the temperature at around 70°C. Oxygen is bubbled in near the bottom of the autoclave, and the reaction, obviously by a different mechanism from that operating in the gas phase employed in the other processes described, goes rather slowly but practically to completion, producing acid of 98% HNO_3.

The Bamag-Meguin process is reported to be widely used abroad. Two of the largest units anywhere, each rated at 440 tons/day, located in Silesia, employ this process. No installations using the Bamag-Meguin process have been reported in the United States.

Cost studies, naturally of a very high-spot nature, carried out as process design assignments by some of my students, show a relatively high total investment when you include the plant for producing the oxygen. This is true even after allowing for the cost of the equipment required to concentrate the nitric acid to 95% as described in the next chapter, when produced at 60% by the Du Pont pressure process.

The Arc Process

The processes so far treated may be called competitive. It is well to mention some of the alternative processes, not starting with ammonia, that have from time to time been proposed for the production of nitric acid, even though they have been tried in the economic balance and found wanting.

The first is the old arc process for nitrogen fixation. Great hopes were held for this process at one time, based as it was on nothing but air and electric power. Air costs nothing, but the consumption of electric power turned out to be so great, per unit of acid produced, that even adjacent to a source of cheap electricity from water power the process could not compete after synthetic ammonia became available.

Cavendish, you remember, had discovered the process back in 1784. The reaction occurs every day. It takes place in lightning flashes; and maybe you have noticed the odor of nitrogen oxides (along with that of ozone) if you have watched someone doing electric welding (with suitable eye protection, I hope), and have realized the importance of adequate ventilation in doing such work.

The process for the purpose of producing nitric acid depends, as you know, on the attainment of a very high temperature, where a moderate concentration of nitric oxide can exist in equilibrium with elemental oxygen and nitrogen, followed by sufficiently rapid cooling of the gases to a temperature where the rate of decomposition of the nitric oxide is negligible.

The first recorded commercial trial took place in America, naturally at Niagara Falls, where "cheap" electric power was available. The furnace as proposed by the inventors, C. S. Bradley and R. Lovejoy, employed high-voltage direct current mechanically interrupted to give a rapid series of arcs. The power consumption was about 9 kw hr/lb of HNO_3. The experimental unit, built in 1902, encountered numerous difficulties and was shut down in 1904.

A contemporaneous development in Norway achieved, for a time, commercial success. Birkeland, a professor of physics at the University of Christiana (now Oslo), knowing that an arc would move at right angles to a magnetic field, conceived the idea of interposing an alternating current arc between the poles of an electromagnet. The magnetic field causes the arc to spread in semicircular form, first in one direction and then in the other, according to the direction of the current through the arc. The alternations are so rapid that the arc appears as a thin disk of flame at right angles to the poles of the magnet.

In collaboration with a Norwegian engineer named Eyde, Birkeland in 1903 devised a furnace based on this principle, with water-cooled copper tubes for electrodes. The furnace was of the form of a flattened cylinder 6 inches deep and 6 feet in diameter, lined with firebrick, through which the air was passed, from center to circumference. The air now containing about 1.2% nitric oxide left the furnace at about 1000°C, and passed through a waste-heat boiler and through coolers made of aluminum, to oxidation tanks and then to an absorption system. As mentioned earlier, this took the form of towers constructed of granite blocks and packed with lumps of quartz. Acid of about 30% strength was produced. Most of it was reacted with limestone to produce calcium nitrate, which was sold as a fertilizer, under the name of Norwegian saltpeter.

In Germany, a parallel study was going on at the B.A.S.F., where O. Schönherr in 1905 discovered that a direct-current arc could be stabilized by confining it in a spirally moving stream of air. With his colleague, Hessberger, he worked out the design of large furnaces based on this observation, drawing up to 1000 hp (746 kw), with arc lengths up to 23 feet. Air, preheated by exchange with the heated gases, passed through holes bored tangentially in the walls of the furnace tube proper. One electrode was water-cooled copper, the other an iron rod, which wore away and was replaced from time to time.

The recovery system for oxides of nitrogen was essentially the same as worked out for the Norwegian process.

The water-power resources in Germany were not considered abundant enough to justify the erection of a plant there, and the large-scale exploitation of the process was moved to Norway, where the first plant was built in 1907. A larger plant was built in 1911, under a joint agreement with the Norsk Hydro (Birkeland-Eyde) firm.

The German concern, realizing by then the potentialities of the Haber ammonia synthesis process, soon withdrew from the venture, leaving it in Norwegian hands. Actually financed from French sources, the Norsk Hydro supplied some nitric acid and nitrates to the Allies during World War I, and a plant using their process was installed in the French Pyrenees.

The large Norwegian plant was shut down a few years later, and by 1928 was replaced by a synthetic ammonia plant using electrolytic hydrogen. (The French plant was likewise replaced.)

There were other processes. A German process, invented by H. and G. Pauling (brothers), employing an arc blown by an air current between diverging electrodes into a fan shape, was installed in Austria, in Italy, in France, and even in the United States. A large installation was finally made in Germany during World War I, in which NO_2 was recovered by refrigeration, using toluene as the refrigerant liquid. It was destroyed in 1917 by an explosion and was never rebuilt. A plant in Switzerland, likewise employing refrigeration, with benzene as the refrigerant liquid, blew up on a day when, believe it or not, the temperature reached 122°F. It was never rebuilt. There was a rather small arc-process plant making sodium nitrite which operated for about ten years at La Grande, Washington. When it burned down in 1927, it was not rebuilt. At the present time, therefore, the arc process remains as of only historical interest, though its independence of raw materials other than air and water causes people from time to time to re-examine its possibilities.

If you have heard of the work being done on what are called arc plasmas, you might begin to wonder whether this might not be something that would give a different aspect to the arc process for direct fixation of nitrogen from the air. The supply of additional current to an arc beyond that necessary to start it results in conduction through an ionized medium and the generation in it of fantastically high temperatures. It is certainly possible, thermodynamically, to obtain at 4000 or 5000°C concentrations of nitric oxide in air as high as those resulting from the combustion of ammonia, and if they could be retained by extremely rapid cooling, the recovery of nitric acid from the gases would not present any greater problem than with ammonia oxidation. But the consumption of electric power would still be high, and as far as I know, nothing of commercial interest has yet come out of this line of study, as intriguing as it sounds.

Fixation by Combustion

The apparent simplicity of the direct high-temperature production of nitric oxide from the air has over the years attracted other proposals. It is known that traces of nitric oxide are produced in internal-combustion engines, a contributing factor in the Southern California smog problem.

Attempts have been made to capitalize on this observation. One such, known as the Häusser process, running on coal gas, reached a fair-sized experimental plant trial in Germany just before World War I. It employed "bombs" of about $2\frac{1}{2}$-cu ft capacity, and produced about 1000 lb/day of HNO_3, in dilute concentration, from exhaust gases containing about 0.5% NO by volume. Cooling was obtained by rapid expansion. The inventor, describing his process in an English journal[4] in 1922, still had great hopes for it, especially if the absorption of the oxides were conducted under 3 atm pressure in stainless steel equipment. But nothing further has been heard of it.

Nitric oxide appears in trace quantities, indeed, in ordinary combustion gases, the proportion depending on the temperature attained and on the rate of cooling of the hot gases.

The "Wisconsin" process represents an attempt, based on a suggestion of F. G. Cottrell (inventor of the Cottrell dust precipitator, and founder of The Research Corporation), to achieve high temperatures by combustion, for example, of natural gas with air preheated in one of a pair of beds of refractory pebbles, used alternately, the other bed used for cooling of the product gases. The process was carried to a large-scale trial under the guidance of Professor Farrington Daniels, with financial support from the University of Wisconsin Alumni Research Foundation, and later from the U.S. Army.

Development of a stable refractory was the first goal sought. Temperatures of 2100°C must be reached, and the product gas must be cooled down to 1500°C at the rate of 20,000°C/sec.

Concentrations of 1.5% nitric oxide by volume were achieved. Such low concentrations would, you know by now, require a very

[4] *Journal of the Society of Chemical Industry* (London), **41**, 253–259 (1922).

large absorption system. Adsorption of the nitric oxide and its catalytic oxidation on silica gel, a porous form of dehydrated silicic acid with a high surface area, was therefore adopted.

A pilot unit was built at the Sunflower Ordnance Works near Lawrence, Kansas, in 1953, capable of producing 40 tons/day of HNO_3, and some 2400 tons were produced. The test was thought to have been a technical success, though the magnesium oxide refractory employed had a rather short life. But it was finally judged not economically competitive, and no further work has been done on it.

Other Processes

There is research being done on the generation of electric power directly from combustion gases by devices that would utilize the phenomena of what is called by a long name, magnetohydrodynamics (MHD). If this ever becomes practical, nitrogen fixation might be a by-product, since extremely high flame temperatures are required.

Solar energy has been proposed as still another source of thermal energy for the direct production of nitrogen oxides from the air. What could be cheaper, or presumably more plentiful, than air and sunshine? Unfortunately, the results of experiments in 1951–1952 at the solar energy research station at Mt. Louis in the French Pyrenees, have been disappointing. Substantial percentages of nitric oxide were obtained all right, in air passed over a thorium dioxide refractory heated at the focus of a solar reflector and then quickly cooled, but the rate of production per square foot of primary mirror surface was too low to be attractive for large-scale application.

What other sources or forms of energy could be brought to bear on the stable nitrogen molecule and cause it to react with oxygen? There is photochemical production of a small, equilibrium concentration of nitric oxide in the upper atmosphere by ultraviolet radiation from the sun. A strong ultraviolet source will indeed produce some nitric oxide (and ozone), which you may have also perceived around a quartz mercury vapor lamp. But the "quantum yield" (the number of molecules caused to combine per unit of energy expended) is far too low to be interesting economically.

You might begin to wonder about the possibility of using the high density of gamma radiation available in a nuclear reactor while in operation to produce power (or to produce plutonium). This has been tried also at the Brookhaven National Laboratory and elsewhere, and here again, while considerable nitric oxide was obtained upon irradiation of air under a variety of conditions, the yield per unit of thermal energy consumed was not sufficient to justify consideration of a nuclear reactor for commercial production of nitric acid.

At the risk, then, of being proved wrong, as Haber was in 1905 when he was discouraged over the prospects for direct synthesis of ammonia, I will say that no process is now in sight that is likely to displace ammonia oxidation as the source of nitric acid.

CHAPTER SIX

METHODS FOR CONCENTRATING NITRIC ACID

You have been reading in earlier chapters that many uses for nitric acid require that it be nearly anhydrous, not the 68% acid of the laboratory stockroom nor the 60% strength usually obtained in the ammonia oxidation process. I now propose to tell you how the principles of physical chemistry are applied in the industrial solution of this problem.

Constant-Boiling Mixture of HNO_3 and Water

That it is a problem you saw in Chapter Two, where the physicochemical data for vapor compositions in equilibrium with aqueous solutions of nitric acid are given in Figure 2.3. It is not just a question of boiling off a more volatile component, leaving the other behind. This works, but only up to a concentration of 68.4% (by weight). At that point the vapor has the same composition as the boiling liquid and comes over unchanged. This is called, from the meaning of the Greek roots, an "azeotrope," a composition unchanged on boiling.

What then to do if you start with a concentration of 60% and want to get a product of 95% concentration?

A physical chemist faced with this problem, remembering the influence of temperature on equilibrium constants, might be tempted to try the effect of a higher or lower temperature, at a corresponding higher or lower pressure, on the composition of the azeotrope. Sure enough, there is an effect. Under a pressure of 150 mm the constant-boiling mixture has a composition of only 66.0% HNO_3, instead of 68.4% at 760 mm. It is theoretically possible, then, to distill a 60% mixture at atmospheric pressure, taking water off the top of the rectifying column and 68.4% acid off the bottom. You would then send this to a second column, operating under vacuum, at 150-mm pressure, and obtain 99.5% acid from the top of this column, and 66.0% from the bottom, boiling point about 77°C at this pressure. This more dilute acid you would send back to the first column and feed it at a point where this concentration was being obtained in going from 60% to 68.4%. It does not take much skill in solving algebraic equations, however, to figure out that you would have to circulate between 12 and 13 lb of HNO_3, 18 or 19 lb of azeotropic mixture between the two columns for every pound of HNO_3 obtained as 99.5%. Distillation columns do not make a separation without reflux returned from the condenser, and it would probably require boiling and condensing three or four times this much acid in each of the two columns. That looks like an uneconomic undertaking, though a German concern (Bamag) thought enough of it to take out a patent, somewhat belatedly, in 1959. The facts had been known for years.

A process variation also employing distillation has been thought up by an American firm (Hercules). This process capitalizes on the temperature effect on the equilibria involved in the production of HNO_3 from NO_2 and water. The constant-boiling mixture (68%) is reacted at a low temperature, say 0°C, with 100% nitrogen peroxide, to produce acid of about 90% HNO_3 by weight, which is then distilled to get 99% acid and the constant-boiling mixture. The nitric oxide produced in the reaction with NO_2 is then passed at a high temperature, say 100°C, through the constant-boiling mixture, regenerating nitrogen peroxide and yielding a lower-strength acid. This more dilute acid is then distilled to remove water overhead and give another portion of constant-boiling mixture.

It is not quite so easy to figure this case, but is apparent that there would be a lot of material recycled in this process and a lot of heating and cooling, as well as boiling and condensing. So it remains to be seen if it will prove economically attractive.

What then remains? It does not take much originality to propose neutralizing the acid with caustic soda (sodium hydroxide, NaOH), or more cheaply with "soda ash" (sodium carbonate, Na_2CO_3), and evaporate the resulting sodium nitrate to dryness and treat it by the old process with sulfuric acid. It would be expensive of course, and a backward step in technology, since you are already starting with nitric acid. Nevertheless, it is recorded that this procedure was adopted for a time in Germany during World War I until improved methods were developed.

A chemist with more originality will propose under such circumstances to take recourse to some dehydrating agent from which the water can be separated in a subsequent operation.

Concentration with Aid of Sulfuric Acid

You have probably already guessed that sulfuric acid was the first dehydrating agent to be tried, since it was already employed in nitric acid technology.

Sulfuric acid was used not only in the old method of producing nitric acid from sodium nitrate, but sulfuric acid was generally required to be mixed with nitric in nitration operations. It was therefore ready at hand when it was necessary to concentrate the weak acid obtained in the absorption towers that served to recover oxides resulting from the decomposition of acid in the niter pots, and from the denitration of spent acids from the manufacture of nitroglycerin or nitrocotton. Such a procedure had been well established by the time ammonia oxidation became commercial, and it was soon adopted as a means of concentrating the acid produced in this process. It is still widely practiced.

Elaborate methods of physical chemistry can be used to determine in what manner sulfuric acid combines with water, and preferentially with nitric acid, when all three are mixed. It is enough for us to know that the composition of the vapors in equilibrium with such mixtures is strikingly different from that in the absence of sulfuric acid. (It should be noted at once that at the temperatures needed to boil nitric acid out of these mixtures, there is a negligible vapor pressure of H_2SO_4, and the vapors can be considered as containing only HNO_3 and water.)

A convenient way of representing equilibrium relationships in a ternary system (a system of three components) is afforded by the triangular diagram attributed to Willard Gibbs, the Yale scientist who formulated the "phase rule." On such a diagram the coordinates of a point, read as the distance to the apex, in each direction, as a fraction of the height of the equilateral triangle, will add up to unity, as you could prove to yourself by geometry.

On the diagram reproduced as Figure 6.1, for instance, at a point at the intersection of the line slanting upward to the right starting from 20% H_2SO_4 with the horizontal line terminating at 40% HNO_3, we find that we also intersect the line slanting downward to the right starting from 40% H_2O; this point then, represents the composition 20% H_2SO_4, 40% HNO_3, and 40% H_2O.

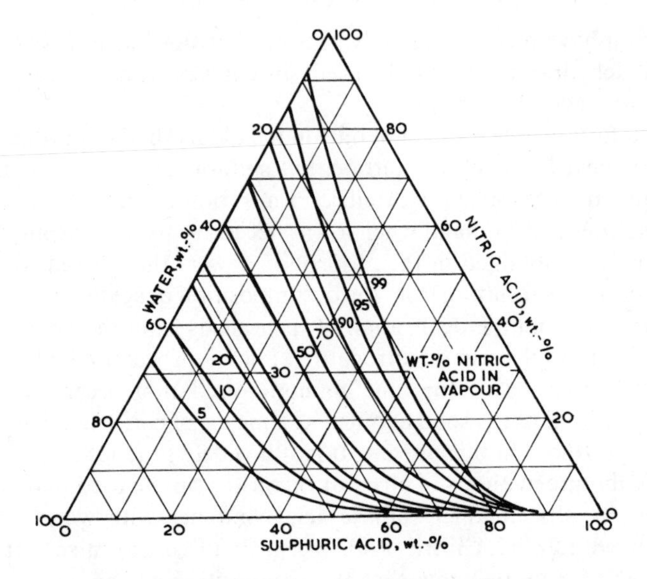

Figure 6.1 Nitric acid—sulfuric acid—water system.

From S. R. M. Ellis and J. R. Thwaites, *Journal of Applied Chemistry (London)*, *7*, 157 (1957), with permission.

It would be understood here as giving the composition of a liquid phase. The curved lines on the diagram show the composition of the vapor in equilibrium with the liquid. In this case, for example, it shows very close to 50% HNO_3. The liquid can be considered as made up by taking 80 parts of a liquid consisting of equal parts of HNO_3 and water, and adding 20 parts of H_2SO_4. You see that no enrichment of the vapor results from adding only this amount of sulfuric, though it is definitely richer in nitric than the vapors of 50% nitric acid, which you can estimate by reading along the left-hand side of the triangle, as between 20% and 30%, maybe 23%.

If we add more sulfuric acid, you will see that the composition on boiling would be definitely above that of the azeotrope. If to 60 parts of 50% acid we add 40 parts of sulfuric, giving a liquid composition of 40—30—30, H_2SO_4—HNO_3—H_2O, the vapor composition would be about 90% HNO_3 by weight. And if to 50 parts we add 50 parts of sulfuric, giving a liquid of 50—25—25, the composition of the vapor is already above 95%. Simple condensation of these vapors would give directly an acid of usable strength for nitrations.

Naturally, removal of nitric acid upon boiling, without much water along with it, would leave a liquid with more water in proportion to the sulfuric, and less nitric, which on continued boiling would give off a vapor of lower nitric concentration.

A useful feature of the triangular diagram is that changes in composition can be readily followed graphically. Removal of HNO_3 would result in a composition moving away from the HNO_3 apex (at the top of the diagram) by a distance proportionate to the fraction removed, as you could prove by geometry. If you had the patience, then you could, by taking successive small increments of vaporization, calculate the changing composition of the distillate and of the residual acid.

Suffice it to say for our purpose here, that such a batchwise vaporization process, starting with a mixture of say 50—25—25 composition, H_2SO_4—HNO_3—H_2O, would give a large fraction of distillate of 90% strength or over, and a smaller fraction of lower strength which would have to be reworked, plus a residual acid containing sulfuric acid and water, with very little nitric. You can see that right down close to the bottom of the diagram the nitric content of the vapor is greater than that of the liquid.

Without such physicochemical data to guide them, but only their observation of what went on, the industrial chemists of an earlier day adopted just such a process to reconcentrate to usable strength the acid obtained upon absorption of the nitrogen oxides escaping from the condensers of the nitric acid retorts, and to recover nitric acid from the spent acids from nitration of glycerin or cellulose. These spent acids contained the water given off in the nitration (esterification) reaction, but still had too much nitric remaining to be thrown away, or to be salable for most purposes to which such lower-strength sulfuric acid could be put, not to mention remaining nitrocompounds, which made them dangerous to handle.

With sulfuric acid already on hand for making nitric acid from sodium nitrate, and niter pots, as the nitric acid retorts were called, it was natural that the practice grew of simply charging a retort with a mixture of weak nitric and strong sulfuric, or spent acid made up to the same proportions, and running a distillation the same way as nitrate of soda was treated to give nitric acid, as described in Chapter One.

The process worked, and persisted for many years, though the life of the retorts in this service was even shorter than in making acid from nitrate, and the cost of fuel for firing the retort was rather high.

A little reflection on the part of a chemical engineer leads to the proposal that a fractionating column, a device for countercurrent contacting of a liquid and a vapor stream, would accomplish the desired result of concentrating the nitric acid much more effectively and economically. He would feed the mixture of nitric and sulfuric and water, preferably hot, almost ready to boil, continuously to the top of a fractionating column, and he would expect to get vapor of 95% or stronger going to the condenser constantly. The residual acid mixture, descending the column, would in successive stages meet the vapor supplied by boiling the acid leaving the bottom of the column, which, if a sufficient number of contacting stages had been provided, would have been almost completely stripped of its nitric content, on account of the high relative volatility of the latter.

The course of such a "stripping" treatment, as the chemical engineer calls it, can be followed even more easily than before on the triangular diagram, and even the required number of theoretical stages can be estimated without much difficulty since steady-state conditions can be assumed. Only a small number are necessary, as a matter of fact.

The use of a stripping column operated after this fashion was indeed described in 1891 in the classic treatise of the German industrial chemist Lunge. Materials of construction were a problem, and the process did not displace the older retort process.

The stripping column for concentrating nitric acid did not come into general use until after the issuance of a patent in 1912 (in the United States) to Harry Pauling, the co-inventor of one of the arc processes for nitrogen fixation described in Chapter Five.

The stripping column process was adopted by the Du Pont Company during World War I for recovery of nitric acid from the large quantities of spent acid from the production of nitrocotton for the supply of the United States and our Allies. A subsequent long-drawn-out suit by the American owners of the Pauling patent was expensive for the Du Pont Company, though the settlement was short of the damages claimed by the plaintiff. I cite this to bring out the bearing which patents can have on industrial processes, and the desirability of protecting your interest or that of your employer by securing patents on anything of value by which you believe you have advanced the art, and at the same time by taking account of the risks you run if you infringe an existing patent, whether by deliberate choice or inadvertently.

It was brought into the testimony that the cost of concentrating nitric acid by the tower process was less than half that of the old retort process.

Upon the introduction of ammonia oxidation as the means of producing nitric acid, the Pauling process, with modifications introduced by the Du Pont chemical engineer, F. C. Zeisberg, so that it came to be known as the Zeisberg process, was widely and almost universally adopted. It remains in use at the present time.

The modifications introduced by Zeisberg were the use of a boiler to generate the vapors for supply to the bottom of the column, instead of direct injection of steam, which dilutes unnecessarily the residual sulfuric acid; and the preheating of the feed, which also cuts down the steam requirement.

Constructional problems were overcome by improved design and methods of connecting up equipment made of high-silicon iron (Duriron) castings, the material most resistant to the combination of hot nitric and sulfuric acids. A unit rated at 150 tons/day of HNO_3 is illustrated in Figure 6.2. You can see the Duriron jacketed S-bends that serve as condensers and boilers.

The cost of the process for concentrating nitric acid by use of sulfuric acid depends partly on the cost of reconcentrating the

Figure 6.2 Nitric acid concentrating plant, using sulfuric acid.
Courtesy Hercules Inc., Wilmington, Delaware.

residual sulfuric or otherwise disposing of it. It would be out of place here to discuss methods of concentrating sulfuric acid in any detail. Let me say only that it is done by either of two methods. One is by means of vacuum evaporation, to reduce the boiling point to where steam can be used to supply the heat. In the other, combustion gases are bubbled through a pool of the acid, evaporating water at a fairly low partial pressure. Construction problems arise in the former, and fume problems in the latter, but both processes are operable.

Reconcentration can be avoided if nearby there is a point of consumption of acid of relatively low strength, say 78%, as for example in the acidulation of phosphate rock to make "superphosphate" for fertilizer use, and if there is also nearby (because freight charges would be too high if great distances were involved) a sulfuric acid plant producing, as is customary, acid of 98% or greater (equivalent) concentration. Under such circumstances, it may be cheaper to pay the premium for the high-strength sulfuric acid over what can be realized from the sale of the diluted residual acid, rather than to go to the trouble and expense of running a sulfuric acid concentrator. To such an extent is the business end of chemical manufacturing related to the technology.

Sulfuric acid is of course employed on plants where nitration operations are carried out. Handling of it therefore poses no unfamiliar problems at such locations. Handling of spent acid from nitrations naturally involves sulfuric acid. Where nitric acid is used for other purposes, however, for example, for oxidations, other dehydrating agents might be considered.

Such compounds as calcium nitrate immediately come to the chemist's mind. Calcium chloride is used as a drying agent in desiccators, and why should anhydrous calcium nitrate not make a good substance for dehydrating dilute nitric acid?

Concentration with Aid of Magnesium Nitrate

It has in fact long been proposed. There was a patent taken out in 1907 for a process by which dilute acid would be added to dehydrated calcium nitrate and the mixture distilled (as in the retort process for concentration using sulfuric acid). In 1913, at the Du Pont Experimental Station, F. C. Zeisberg made determinations of the strength of vapors coming off of mixtures containing calcium nitrate. I made additional measurements myself, in 1929, and found that in the presence of enough $Ca(NO_3)_2$ to form a saturated solution at the boiling point with the water content of ammonia oxidation acid, the composition of the vapors was well above that of the azeotrope. A short rectifying column above the stripping column, with a very moderate reflux, would then suffice to give 99% acid if desired. The process looked entirely feasible, indeed economically attractive. But the fact that sulfuric acid could be handled as a liquid—though with recognized corrosion problems—as against the probable necessity of drying the calcium nitrate to solid form, made the continued use of the sulfuric acid route appear justified.

No active consideration was given at that time to magnesium nitrate, supposing that it would be more subject to decomposition upon dehydration than calcium nitrate.

Recurrent interest, nevertheless, was shown from time to time. During World War II a process was worked out at one of the ordnance plants on a pilot scale employing magnesium nitrate. United States Patents covering magnesium nitrate concentrating processes were subsequently issued and assigned to Chemical Construction Co. (1949 and 1958), Hercules Powder Company (1955), and Eastman Kodak Company (1963). Each of these covered different arrangements of rectifying and stripping columns and of reboiler and evaporator. Though they claim any alkaline earth nitrate, all specify the use of magnesium nitrate.

The process using magnesium nitrate has been developed on a commercial scale by Hercules.[1] A unit installed at one of their plants is shown in Figure 6.3. The way it operates can be readily understood by reference to the simplified flowsheet, Figure 6.4.

Acid from an ammonia oxidation unit of about 60% strength is fed at the top of the stripping (lower) section of a dehydrating

[1] *Chemical and Engineering News*, 40–41 (June 9, 1958); R. J. Bechtel, paper delivered at meeting of the American Institute of Chemical Engineers, February, 1961.

Figure 6.3 Nitric acid concentrating plant, using magnesium nitrate.
Courtesy Hercules Inc., Wilmington, Delaware.

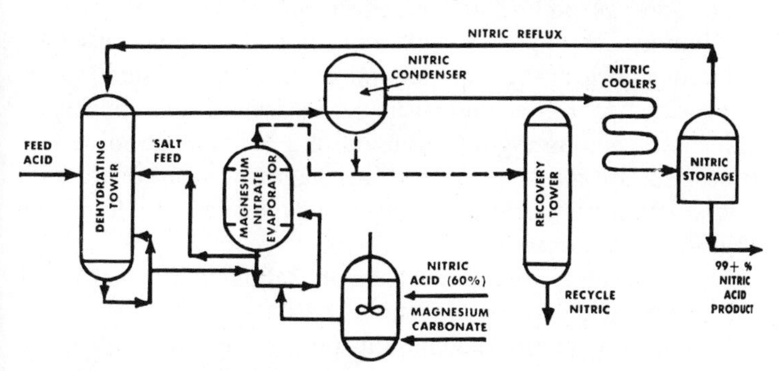

Figure 6.4 Flowsheet of magnesium nitrate concentrating process.
From *Chemical and Engineering News, 36,* 40, June 9, 1958, with permission.

tower. Magnesium nitrate solution, 72% $Mg(NO_3)_2$ by weight, at 140°C, is fed at the same level (or it may be premixed). Vapors from this mixture of salt solution and weak nitric acid feed contain about 87% HNO_3 and 13% water. These vapors rise into the bottom of the rectifying (upper) section of the tower. Only a short section is required, and only a moderate fraction of the overhead condensate has to be returned to obtain a concentrated nitric acid product, 99.0 to 99.5% HNO_3. The bottom product from the rectifying section, about 75% HNO_3, 25% water, flows down into the stripping section, mixing with the feed acid and the dehydrating solution. In the stripping section, supplied with steam at the base, obtained by boiling the residual $Mg(NO_3)_2$ solution (in a reboiler, not shown on the flowsheet), the nitric acid is "stripped" from the mixture fed at the top, and the residual solution, boiling at 170° to 180°C, contains about 68% $Mg(NO_3)_2$, with less than 0.1% HNO_3, close to the equilibrium hydrolysis value.

The necessary quantity of residual solution is taken to a vacuum evaporator where it is reconcentrated to 72%. A little slide-rule work will show that it takes the return of close to 700 lb of 72% solution to dehydrate 100 lb of feed containing 60% HNO_3 if the $Mg(NO_3)_2$ content is reduced to only 68% by weight.

These solutions can be handled easily in stainless steel equipment, and the boiling and reconcentration can be done with steam at moderate pressure (400 lb/sq in.). Aluminum is used for the strong nitric acid condenser and storage tanks. A small tower is provided to recover nitrogen oxides leaving the condenser. Salt losses, amounting to a fraction of a pound per ton of acid processed, are made up by treating magnesium carbonate with feed-strength nitric acid and then passing the solution through the $Mg(NO_3)_2$ concentrator.

Operating costs are claimed to be only about half as high as for the sulfuric acid process, and capital costs one-third less.

So it turns out that magnesium nitrate does not have to be reduced to dryness to be effective as a dehydrating agent, and therefore can be handled as a liquid; and moreover it is not subject, as I feared it would be, to excessive decomposition on reconcentration.

The moral I would draw from this story is one that I could quote from the Reverend Mr. Milner's paper of 1789: "It is best ... in such matters to trust as little as possible to conjectures, and to bring every opinion to the test of experiment."

CHAPTER SEVEN

USES AND APPLICATIONS

Uses for nitric acid have been greatly diversified and market demands have been enormously expanded since the properties of *aqua fortis* were first recognized by the alchemists.

Dissolving Metals

The use the alchemists made of nitric acid, to bring metals into solution, still remains. One important example, silver nitrate, produced by dissolving silver in nitric acid, is used in making emulsions for photographic films and papers. Along with hydrochloric acid, as *aqua regia*, it is required in extracting and refining the noble metals, gold and platinum. Photoengravers use it to dissolve copper from plates to be used in printing, in the areas where the "resist" has not been hardened by exposure to light, so leaving those areas at their original height and capable of taking ink from the rollers and transferring it to the paper.

An important, though still not very large, application is in the field of atomic energy. Processes making use of atomic (more properly, nuclear) power depend, so far, upon uranium. The nitrate salt of uranium is soluble in organic liquids, and uranium is purified by taking advantage of this solubility property, in contrast to the insolubility of the nitrate salts of most other elements. It can therefore be extracted from a water solution, leaving contaminants behind. This property is useful in preparing pure uranium for use as a fuel in a nuclear reactor; it is even more important in recovery of plutonium (which has the same property, in a suitable state of oxidation) and uranium from the mixture of radioactive fission products that result from the nuclear reaction.

There is a rule that "all nitrates are soluble in water," which is so nearly universal that it is difficult to find a reagent that will form a precipitate with it, though there is one. It is called "nitron," a complex organic compound with the empirical composition $C_{20}H_{16}N_4$. It is not so well known that certain metallic nitrates other than uranium are also soluble in organic solvents, such as ether (diethyl ether). (A more effective solvent, used in the nuclear energy industry, is tributyl phosphate "TBP," diluted with kerosine.) Most other nitrates are not soluble in these solvents. The explanation is that the latter, of the sodium nitrate and calcium nitrate type, are ionic salts, the metal having lost an electron to the nitrate ion. The former are more properly designated as "nitrato-compounds," with covalent chemical bonds and a structure of the type

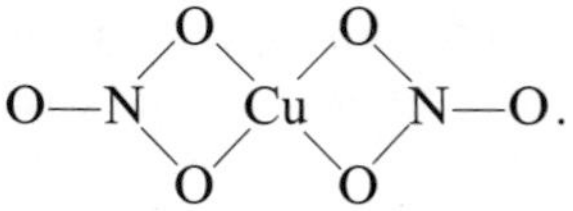

```
      O     O
     / \   / \
O—N     Cu    N—O.
     \ /   \ /
      O     O
```

The hydrate of copper nitrate for example, is soluble in TBP, just as is that of uranyl nitrate.

In addition to its traditional role as a reagent to dissolve metals, nitric acid is in demand to fill at least four other functions. It serves as an acid, reacting with alkalies to form salts, and with alcohols (or organic compounds containing the OH hydroxyl group) to form esters; it reacts with hydrocarbons as a substituent to form nitrocompounds; and it serves as an oxidizing agent.

Ammonium Nitrate

The first of these uses has come to be the one for which the largest quantity is required. This is because of the desirable properties possessed by one salt, ammonium nitrate.

Ammonium nitrate has for many years been employed as a constituent of high explosives. Nobel took over a patent of

Björkmann of 1867 covering mixtures of nitroglycerin and ammonium nitrate, and introduced a line of "extra dynamites," variations on which are made to the present day.

Ammonium nitrate alone is not very powerful and is difficult to set off:

$$NH_4NO_3 \rightarrow N_2 + \tfrac{1}{2}O_2 + 2H_2O.$$

As an explosive it has an "oxygen excess," as you see from the equation for its decomposition. As an explosive, therefore, its effectiveness can be improved by incorporating an oxidizible substance in admixture with it. Many such compositions were tried, but the first practical explosive of this type was not developed until about 1935. In this composition the ammonium nitrate was coated with 5% paraffin wax and was packaged in metal cans. It became popular under the name of "Nitramon."

The wax coating served to prevent caking of the ammonium nitrate grains, and the metal can effectively sealed against moisture penetration. Such protection is important because, as a highly soluble salt, ammonium nitrate would pick up moisture from the air and its effectiveness would be reduced.

You might at this same time acquaint yourself with the fact that the salt undergoes several changes in crystal structure with changing temperature, like those that sulfur undergoes, which you may have observed in the laboratory (see Table 7.1). You will note that one of these, involving a volume change of 3.6%, occurs at a temperature of around 90°F, which would be hard to avoid on storage. If the grains of the crystalline material were in direct contact during a succession of cycles through this temperature, you can imagine that the mass might easily cake together. This effect is in addition to the customary caking tendency shown by ordinary hygroscopic salts such as sodium nitrate or even sodium chloride.

These factors limited its use as a constituent of military high explosives, except in admixture with other explosive compounds, such as TNT (trinitrotoluene); this mixture, known as Amatol, was in great demand during World War I. You were reading in Chapter Three about the construction of factories for the production of ammonium nitrate in the United States in 1918.

TABLE 7.1

Crystal Forms of Ammonium Nitrate

Modification	Temperature Range °C	°F	Crystal System
I	169.6–125.2	337.3–257.4	Cubic
II	125.2– 84.2	257.4–183.6	Tetragonal
III	84.2– 32.1	183.6– 89.8	Rhombic
IV	32.1–(−18)	89.8–(−0.4)	Rhombic (pseudo-tetragonal)
V	Below −18	Below −0.4	Tetragonal

Changes During Crystal Transition

Transitions	Temp. °F	Heat Change	Volume Change
V → IV	−0.4	1.6 cal/g absorbed	2.9% shrinkage
IV → III	89.8	4.7 cal/g absorbed	3.6% expansion
III → II	183.6	4.0 cal/g absorbed	1.3% shrinkage
II → I	257.4	12.6 cal/g absorbed	2.1% expansion

(After *Manual A*-10 "Ammonium Nitrate," Manufacturing Chemists' Association, Washington, D.C., with permission.)

In addition, ammonium nitrate, with 35% nitrogen, has properties that make it eminently desirable as a fertilizer ingredient. Its use was held back for a long time by its tendency to cake during storage. A fertilizer must naturally be free-flowing to be useful in the field. The development of effectively moistureproof multiwall paper bags for shipment and storage and the discovery that caking could be prevented by coating came along about the same time, and made it practical to consider ammonium nitrate for fertilizer. The development of a process for shotting or "prilling," as it is called (in the Cornish dialect nuggets are called "prills") also helped. After World War II there was for a time an excess of nitric acid production capacity, and it was economic to use this surplus plant equipment to supply acid for ammonium nitrate manufacture.

The first anticaking material to be widely used was wax, as found effective in "Nitramon," and large tonnages were produced. Two or three thousand tons were being loaded in two vessels at Texas City, Texas, in 1947, for shipment to Europe, when fire broke out on one of them, resulting in a tremendous explosion;

fire and explosion on the other vessel followed. Four hundred people were killed and more than a thousand injured.

This disaster and an explosion a few months later on a ship about to be unloaded at Brest, France, naturally gave cause for alarm as to the safety of handling ammonium nitrate, and a thorough investigation of these cases and of means for avoiding such dangers was undertaken.

The material involved had been treated with wax as an anticaking agent, albeit only 1%. Unwise measures were employed in attempts to control the fires that got started, measures that are effective with combustible materials that require access of air to continue burning, namely, confining the material and smothering with steam. Ammonium nitrate supplies its own oxygen.

Recommendations for safe packaging and transportation of fertilizer grade ammonium nitrate are given in *Manual Sheet A*-10 put out by the Manufacturing Chemists' Association, and millions of tons of the material have been handled now without mishap. Only inert inorganic materials are now used to prevent caking, generally diatomaceous earth or fine clay. And only copious drenching with water is prescribed in the event of a fire, and provision for free release of any pressure that might tend to build up.

Before the advent of nitric acid via ammonia oxidation, ammonium nitrate was in an emergency produced by double decomposition between ammonium sulfate from by-product coke ovens, and imported sodium nitrate. The process, worked out in England, was practiced in the United States during World War I. A plant at Perryville, Maryland, was said to have turned out as much as 500 tons/day.

At present of course it is made by neutralization of ammonia with nitric acid, as indeed it had generally been for a long time. You can hardly think of a simpler process to carry out:

$$NH_3 + HNO_3 = NH_4NO_3.$$

Indeed it is rather simple, and such variations as are found in industrial practice are directed mostly toward heat economy, and depend on the source of nitric acid available.

The reaction gives out heat to the extent of 26,000 cal/g mole, and if ammonia is supplied as a gas and the nitric acid is sufficiently concentrated, the heat generated is sufficient to evaporate most or all of the water associated with the acid.

A complete description of a typical plant for producing fertilizer grade ammonium nitrate has been published,[1] and a flowsheet is given in Figure 7.1, which you should find almost self-explanatory.

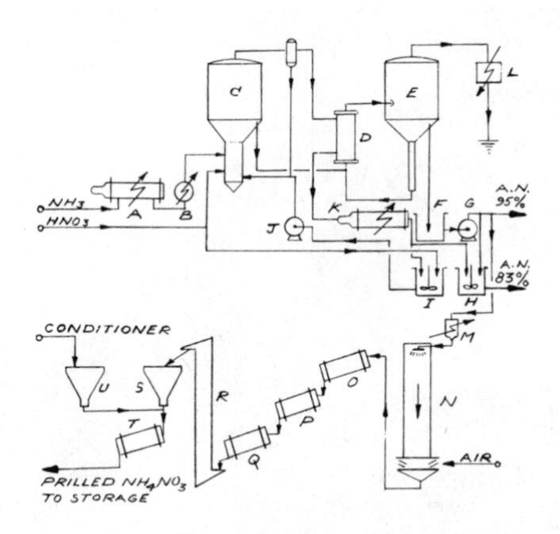

Figure 7.1 Pressure neutralizer concentrator process.

A. Ammonia heater
B. NH_3 superheater
C. Neutralizer
D. Evaporator heater
E. Evaporator
F. 95% A.N. Tank
G. 95% A.N. Pump
H. 83% A.N. Mixer
I. Reclaim tank
J. Reclaim pump
K. NH_3 flash tank
L. Evap'r. condenser
M. Film evaporator
N. Prilling tower
O. Pre-prill dryer
P. Prill dryer
Q. Prill cooler
R. Prill elevator
S. Prill bin
T. Coating drum
U. Conditioner bin

From V. Sauchelli, *Fertilizer Nitrogen: Its Chemistry and Technology*, American Chemical Society Monograph No. 161, p. 236, Reinhold, New York, 1964, with permission.

Briefly, ammonia is vaporized and superheated and bubbled into a continuous neutralizer also supplied with 55% nitric acid. The heat of reaction is sufficient to concentrate the liquor (plus redissolved fines) to about 83%, boiling at a temperature of 130°C. The neutralizer is maintained slightly on the acid side, pH of 1.5 (a measure of hydrogen ion concentration). The hot solution is then fed continuously to a "long-tube falling-film" evaporator,

[1] *Industrial and Engineering Chemistry*, **45**, 496–504 (1953).

operating at 17 in. Hg vacuum (about 0.5 atm), where it is concentrated to 95%. The solution is then run to the prilling tower at 140°C. The tower at the plant described is 100 feet high and 20 feet in diameter. (In other plants, towers of twice or even three or nearly four times this height are provided.) Two such towers are shown in Figure 7.2.

Figure 7.2 Prilling towers for ammonium nitrate.
Courtesy Manufacturing Chemists' Association, Washington, D.C.

The prills are then successively dried and cooled in slanted rotating cylinders through which, in the first, heated, and, in the last, cold air is passed. They are treated with dried diatomaceous earth, a finely divided silica mineral, to the extent of 3.5%. The product is then ready for bagging and shipment.

Most of the nitric acid produced in the United States is used in the manufacture of ammonium nitrate, the proportion being between two-thirds and three-fourths in recent years (1959–1962). In 1962, 3.4 million tons were made, most of it as "fertilizer grade," roughly 90%. It is not easy to ascertain from the published statistics, however, how much "fertilizer grade" material was actually consumed in "do-it-yourself" explosives applications. It came to be observed that in surface mining and quarrying, loose ammonium nitrate with fuel or Diesel oil added was an effective blasting agent. The practice has been worked out in iron mining in the Mesabi Range in Minnesota simply to pour into 9-in. bore holes a bagful of prilled ammonium nitrate, 80 lb, followed by 1 gallon of fuel oil, as many times as required, with finally a charge of dynamite and a blasting cap. The practice is apparently growing and is perhaps becoming more sophisticated.

One variation that has come into large-scale use violates the time-honored admonition to "keep your powder dry" (from the days, of course, of gunpowder). These formulations are made up with water solutions of ammonium nitrate, of high concentration, plus wax and a sensitizer such as TNT, and a gelling agent.[2]

Synthetic Sodium Nitrate

Sodium nitrate, once the principal source of "fixed" nitrogen, has declined in importance, as you have been reading in earlier chapters. It can be, and is, produced by the method mentioned in Chapter Five by supplying sodium carbonate to the final towers in an atmospheric pressure ammonia oxidation plant. Any nitrite formed in the alkali towers is decomposed in the acid tower nearer the entrance for the gases coming from the ammonia converters; the oxides resulting go forward with the unabsorbed oxides into the alkali towers. A flowsheet of the process is given in Figure 7.3. The acid solution of sodium nitrate is finally heated and blown with air; then neutralized with sodium carbonate, settled, evaporated, filtered off, and the salt dried and screened. The production amounted to less than 0.5% of the total of fixed nitrogen in the United States in 1962.

On the market it comes into competition with imported Chilean nitrate. The domestic producers stress the purity of the product; the importers of Chilean nitrate make a point of the "trace elements" their product contains. But at the price one has

[2]M. A. Cook, *Science*, **132**, 1105–1115 (Oct. 21, 1960).

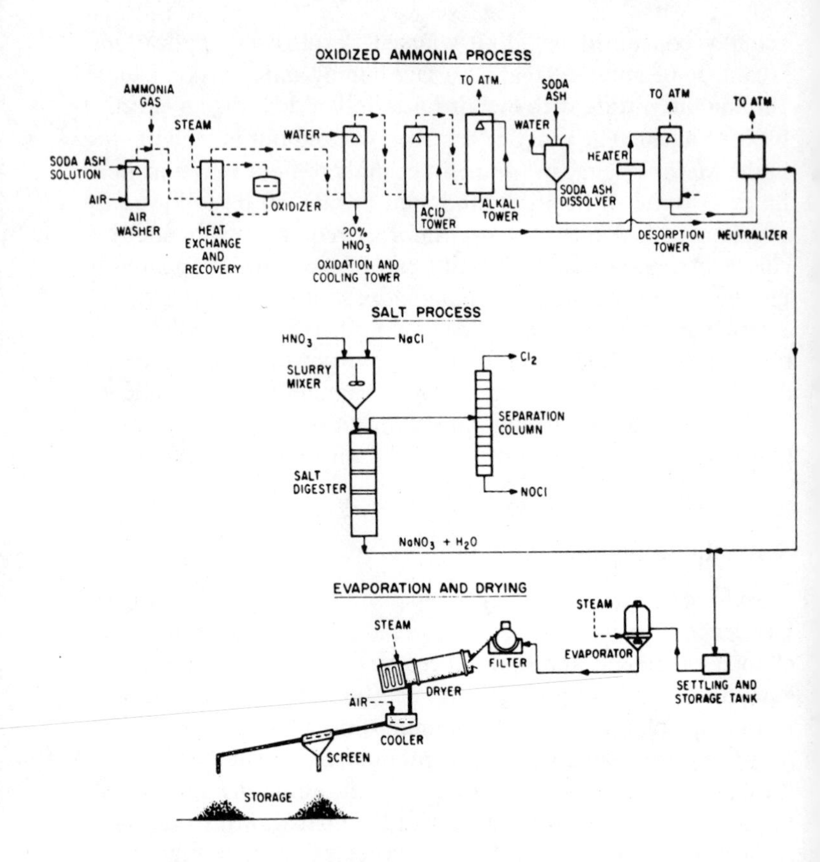

Figure 7.3 Synthetic sodium nitrate process.
From V. Sauchelli, *op. cit.*, p. 335, with permission.

to pay for either (see Table 3.1, p. 65), priced competitively each with respect to the other, at $2.75 per "unit" of nitrogen, it is small wonder that the demand for both is declining. An inference can be drawn from the table while we are looking at it: Anhydrous ammonia is the cheapest thing to use per unit of fixed nitrogen content. As a matter of fact, its use for direct injection into the soil is increasingly rapidly.

For comparison, the current quotation for prilled ammonium nitrate is about $2.00 per unit (20.0 lb) of nitrogen.

An interesting variation of the process for the manufacture of sodium nitrate is shown sketchily on the flowsheet, Figure 7.3. This is the process starting with ordinary salt, sodium chloride, which is the raw material for making sodium carbonate. The process ought to be attractive for that reason, but you will at once realize that the highly corrosive mixtures of nitric and hydrochloric acids would have to be handled.

As developed by Allied Chemical Co. at Hopewell, Virginia, the corrosion problem was met by the device of surrounding the walls and cover of an acidproof brick reaction vessel with an outer mantle of stainless alloy, and filling the space between with a nitrate solution, or simply washing down the walls, with a solution of sodium carbonate. (The base is exposed only to nitrate and can be made of stainless steel.)

The flowsheet shows simply the separation of the nitrosyl chloride-chlorine mixture formed in the reaction,

$$3NaCl+4HNO_3 = NOCl+Cl_2+2H_2O+3NaNO_3,$$

by straightforward distillation, after removing the water from the gases by means of refrigerated 66% HNO_3. Chlorine boils at −34.6°C and nitrosyl chloride at −5.5°C; but the separation is made under pressure where the boiling points are higher.

Originally, the NOCl was reacted with soda ash (sodium carbonate):

$$3NOCl+2Na_2CO_3 = NaNO_3+3NaCl+2CO_2+2NO.$$

This necessitated reworking a good deal of salt (sodium chloride) and nitric oxide and reduced the attractiveness of the process. With the decline in market demand for nitrate of soda, the process was abandoned.

More recently another method of working up the NOCl has been developed, namely, oxidation with oxygen to chlorine and nitrogen dioxide, which puts the process in a different perspective. According to the over-all equation

$$6NaCl+8HNO_3+O_2 = 6NaNO_3+4H_2O+3Cl_2+N_2O_4,$$

you would recover all of the sodium of the salt as nitrate, all of the chloride as chlorine, and one-third of the nitrogen as nitrogen peroxide. With the selection of nitrogen peroxide as a storable oxidizer for rocket propulsion, for example for the Titan II, this may prove to be an economically feasible process for its production.

Other Nitrate Salts

Potassium nitrate, saltpeter or niter, the original source of chemical nitrogen, as you read at the beginning of Chapter One, is made principally by conversion of the chloride, the most abundant mineral form, with sodium nitrate. With nitric acid as easily available now as sodium nitrate, it has become of interest to make nitrate of potash from the chloride by a process like that just described for making sodium nitrate, with chlorine as a by-product. A large plant has been built at Vicksburg, Mississippi, and is reported to be in production, after overcoming anticipated corrosion difficulties. Containing as it does two essential fertilizer elements, nitrate of potash might command a premium as a constituent of concentrated fertilizers.

Barium and strontium nitrates are required in small amounts for pyrotechnics and signal flares. They are produced simply by treating the carbonates with nitric acid and crystallizing.

Magnesium nitrate, as you read in the preceding chapter, is used in one process for concentrating nitric acid. It is produced on the plant as needed.

Calcium nitrate, the principal product of the old arc process plants, is not now produced except as an incidental product in treatment of phosphate rock with nitric acid.

It is hygroscopic, and fertilizer mixtures containing it are subject to caking in storage. Therefore, it is generally converted to some other form, as by ammoniation.

With nitric acid abundantly available, there is an incentive to employ it, instead of sulfuric acid, for the acidulation of phosphate rock because the nitrate radical would contribute to the value of the product as a fertilizer, while sulfate is not (ordinarily) needed by the soil. There is continuing development of processes directed along these lines, in Europe and in the United States. The Tennessee Valley Authority (TVA) is particularly active in this field, outside the scope of this book, except as this promises to afford a growing outlet for nitric acid.

Nitrate Esters

I want to consider next the applications of nitric acid in which the nitrate ion reacts with hydroxyl (OH) groups to form what the organic chemist calls esters. The resulting compounds, containing both oxidizing and reducing groups, constitute potential explosives, and it is as explosives that most of them find their usefulness. This is not a book on explosives; nevertheless, I think it worth while to outline typical processes using nitric acid for their manufacture.

The best-known materials of this nature are ordinarily known as nitroglycerin and nitrocellulose, referred to by these names at the very beginning of Chapter One. The organic chemist objects to these names, preferring glyceryl trinitrate and cellulose nitrate, respectively, since they are actually nitrate esters of polyhydroxy alcohols and not nitrocompounds formed by substitution of hydrogen atoms, like nitrobenzene and nitromethane.

The reaction,

$$ROH + HNO_3 = RNO_3 + H_2O,$$

for the esterification of an alcohol, produces water, and the presence of some material that will combine with the water is generally required to bring the reaction to completion. The customary material is concentrated sulfuric acid, which also serves as a catalyst.

Nitroglycerin

Because of its industrial and historical importance, we shall present here a brief outline of the traditional batch process for the manufacture of nitroglycerin, as given in a standard reference work.[3]

Nitroglycerin is manufactured by adding slowly 1 part by weight of glycerin (approximately 700 lb) to about 4.3 parts of a mixed acid, contained in an iron, steel or lead nitrator. The anhydrous acid contains about 52.7% nitric acid, 49.4% sulfuric acid (equivalent, including dissolved SO_3), and 0.35% nitrosylsulfuric acid. It is stirred constantly by means of compressed air and cooled by means of brine coils, while the temperature is kept at 25°C or less. Should control of the temperature be lost or red fumes evolve, the charge is dumped into a large drowning tank full of water. After addition of the glycerin to the mixed acid is completed, the emulsion of nitroglycerin in water and acid is subjected to additional agitation and cooling until the temperature is about 15°C. The charge then is run into a separating tank where the nitroglycerin forms a supernatant layer containing about 8% nitric and 2% sulfuric acid. Agitation of the nitroglycerin with water at a temperature as high as 43°C (drowning wash) removes most of the dissolved acid. After settling out, the nitroglycerin is given additional washes with water, 2% sodium carbonate solution, and more water until the wash waters are free from alkali and the nitroglycerin is neutral to litmus. The purified nitroglycerin which appears milky because of the moisture content, is transferred to storage tanks in a heated building. Here it rapidly becomes colorless and the moisture content decreases to 0.4% or less. The yield of nitroglycerin is 230 ± 5 parts by weight per 100 parts of glycerin.

The spent acid from the nitration of glycerin contains approximately 10% nitric acid and 73% sulfuric acid. This is subjected to processing for recovery of the nitric acid and concentration of the resulting dilute sulfuric acid.

A flowsheet of the process as described is given in Figure 7.4.

This is essentially the process as employed by Nobel 100 years ago. A continuous process has long been sought, and one or two have been developed.

[3]R. E. Kirk and D. F. Othmer, Eds., *Encyclopedia of Chemical Technology*, 1st Ed., Vol. 6, p. 28, Interscience, New York, 1952.

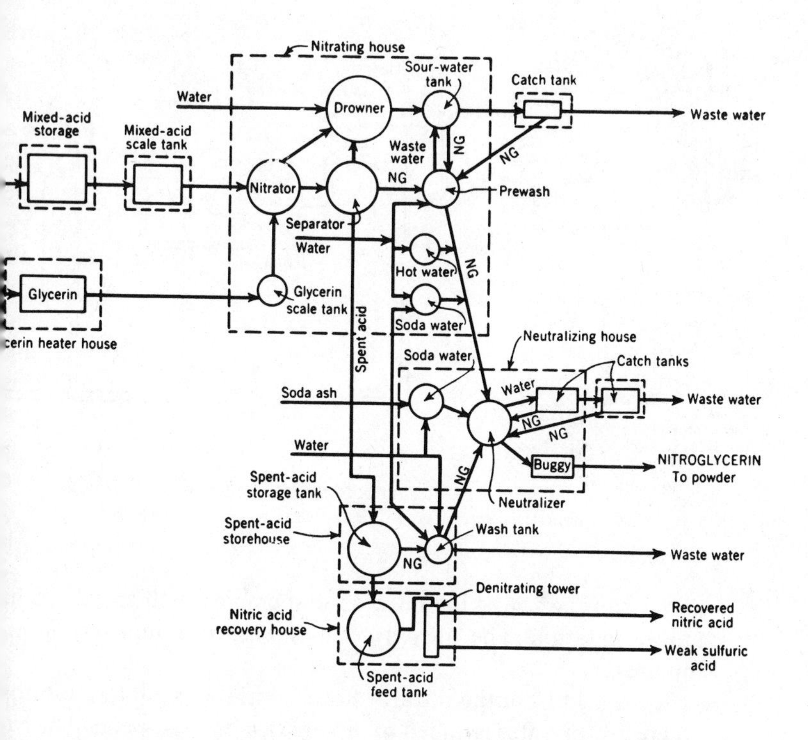

Figure 7.4 Batch process for the manufacture of nitroglycerin.

rom R. E. Kirk and D. T. Othmer, Eds., *Encyclopedia of Chemical Technology*, Vol. 9, p. 324, 1st ed., Interscience, New York, 1952, with permission.

The Biazzi process, patented in 1935, follows the outline shown n Figure 7.5. It offers the advantage of greater safety because of he smaller amount of nitroglycerin in the system at any one time, nly 25 to 75 lb when producing 1000 lb/hr in a 10-gal nitrator. The nitrator has helical cooling coils designed to permit rapid eaction through rapid heat absorption.

In the separator, provision is made for slowly rotating the inter- nediate emulsion layer, with consequent improvement in the reaking of the emulsion, and prevention of local overheating and ormation of "deadspots." The stainless steel washing tanks are of

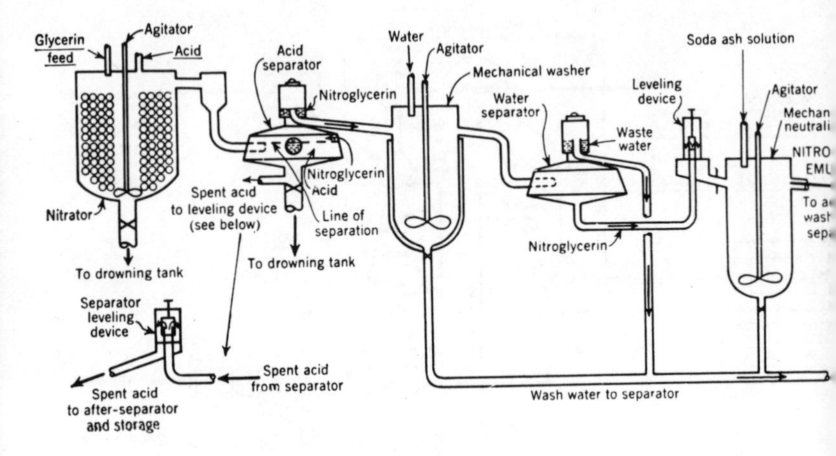

Figure 7.5 Biazzi continuous nitration plant.
From Kirk and Othmer, *op. cit.*, 1st ed., Vol. 9, p. 326, with permission.

about half the size of the nitrator, and employ high-speed mechanical agitation. The nitroglycerin is washed completely in 10 minutes.

Spent acid from the nitrator flows continuously into a dilutor where 2% of water is added to dissolve traces of suspended nitroglycerin. Automatic control devices indicate if temperatures are too high or too low, to cut off feeds if there is a current or belt failure, and to "drown" the nitration mixture if the temperature exceeds a set limit. The nitroglycerin produced by the Biazzi process is claimed to be of unusual purity and stability.[4]

Nitroglycerin is a yellow, oily liquid, density 1.6, freezing point 13.2°C (55.8°F). The liquid is used as such as an explosive only in the extremely hazardous business of putting out oil well fires, as you may have read about from time to time. It has an appreciable vapor pressure at ordinary temperature, and it has sufficient physiological activity to give some people who handle it a persistent headache. Small quantities of a purified grade, adsorbed onto an inert material, are used in medicine, as an agent to dilate the

[4]This outline also based on R. E. Kirk and D. F. Othmer, *op. cit.*

blood vessels, though it has been somewhat superseded in this application by pentaerythritol titranitrate (PETN).

Nitroglycerin, discovered by the Italian, Sobrero, was found to be a powerful explosive, when set off by an initiator, which Alfred Nobel invented, but it proved alarmingly unpredictable. Nobel's second invention made it practical. Contrary to a story current for these 100 years, it came about not by accidental discovery of a pasty mass formed by a leaking container packed in diatomaceous earth, but by methodical search on the part of Nobel, who tried various absorbents, and decided upon "kieselguhr," as this powder is known in German, on account of its low apparent density and chemical inertness. In any event, "it was found that kieselguhr would easily absorb as much as three times its weight of nitroglycerine, that the resulting putty-like substance could be kneaded and packed in cartridges, that it was less sensitive to shock or blows but could nevertheless be exploded with a blasting cap. Nobel called this mixture 'dynamite' . . . and proceeded to patent and manufacture it."[5]

This is why you read in most chemistry books that dynamite is a mixture of nitroglycerin and kieselguhr. Actually, better absorbing materials were soon developed, which either were themselves explosive, such as ammonium nitrate, or contributed explosive power, like sodium nitrate when mixed with wood flour or other combustible material. These were called "dopes," and the resulting explosive "active dynamite."

It was also found that nitrocotton could be gelatinized with nitroglycerin:

"The invention of blasting gelatine, while perhaps directly due to an accident, was the result of Nobel's search for a substance that would retain nitroglycerine better than the absorbents in use, one that would preferably enter into a solid solution with it and at the same time take an active part in the explosion. Guncotton, which had been listed as an absorbent in the patent of 1863, did not give the desired results. But one day Nobel, who had moved to Paris in 1873 and established his laboratory there, hurt his

[5] A. P. Van Gelder and H. Schlatter, *History of the Explosives Industry in America*, pp. 328–329, Columbia University Press, New York, 1927.

finger and used collodion (a solution of nitrocellulose, of lower nitration than guncotton, in a mixture of ether and alcohol) to cover the wound. At two o'clock in the morning, being unable to sleep on account of the pain, he went to his laboratory to try the effect of collodion on nitroglycerin. To his satisfaction he found that after evaporation of the solvent a semi-solid mass resulted. Further experiments showed that gentle heating would make the material tough, plastic, and elastic, and that he could obtain the same result by adding 7 to 10% of collodion nitrocotton directly to nitro-glycerine. Smaller quantities of nitrocotton gave a thinner gelatinized oil which had the property of being retained by suitable active dopes without exudation of the nitroglycerine. He called the plastic material 'blasting gelatin' and the mixtures with dopes 'gelatin dynamites'. Both were patented in 1875."[6]

"Blasting gelatins" have found wide application, particularly for underwater blasting. Under the name of double-base powder, with less "NG" and more "N/C," they find use as propellant explosives and lately as solid propellants for rockets and missiles.

Nitrocellulose

The other principal nitrate ester is nitrocellulose. Discovered by Schoenbein about the same time as nitroglycerin, it did not become commercially useful until means were found for purifying and stabilizing it, again about the same time as the invention of dynamite. Nitrocellulose burns at the surface of a block or "grain" into which it is compressed, very rapidly, it is true, at a rate which makes it useful for propelling projectiles, but it does not detonate if the "grains" are properly shaped. It is the primary reliance for propellant charges in conventional weapons (as opposed to nuclear), and has therefore been an essential material for warfare for all wars in this century. It also has properties that make it (in a less highly nitrated grade) useful in plastics and in coatings. An important plastic was the one called "Celluloid," by its inventor, Hyatt. Celluloid is made by compounding with camphor (originally to substitute for ivory in making billiard balls). The original "Duco" lacquer was also based on nitrocellulose.

[6] A. P. Van Gelder and H. Schlatter, *op. cit.*, p. 339.

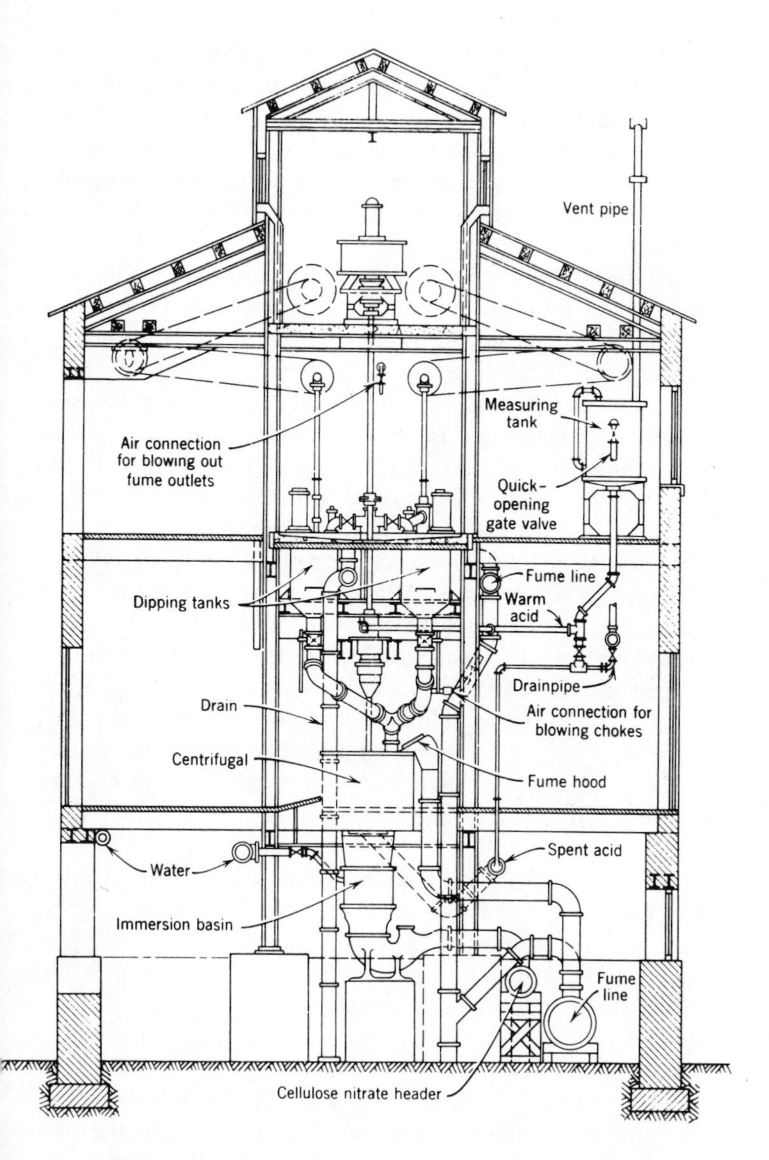

Figure 7.6 Mechanical dipper process for nitrating cellulose.
From Kirk and Othmer, *op. cit.*, 2nd ed., Vol. 4, p. 629, 1964, with permission.

The preferred raw material for nitrocellulose is cotton "linters," the short staple, unsuitable for spinning, that adheres to the cotton seed. Wood pulp can also be purified to give a nitration-grade cellulose suitable for many purposes.

The usual commercial process for the manufacture of cellulose nitrate is the mechanical-dipper process developed by E. I. du Pont de Nemours & Co., Inc. In Figure 7.6, a schematic diagram of a cellulose nitrate unit is shown. Such equipment consists of a measuring tank for the mixed acid; four dipping pots or reactors, in which the cellulose and mixed acids are stirred together; one centrifuge, which removes most of the spent acid from the cellulose nitrate; and an immersion basin in which the centrifuged cellulose nitrate is drowned—that is, rapidly mixed with a large amount of water. About 30 lb of cellulose (chemical cotton or chemical wood pulp, depending upon the product being prepared), containing less than 1% moisture, is added to the dipper containing about 1600 lb of mixed acid. The reaction temperature is controlled by adjusting the temperature of the acid before it is added to the reactor. The four reactors are fed and controlled by one operator. When nitration has been completed, the reaction mixture is dropped into the centrifuge, which removes most of the spent acid. The latter is pumped to a tank where it is brought back to strength for reuse by adding concentrated acid. The centrifuged cellulose nitrate is dropped through the bottom of the centrifuge and drowned in water. It is then pumped as a water slurry to the purification area.

One reason for preparing the cellulose nitrate in comparatively small batches is that it is necessary to centrifuge and drown the cellulose nitrate as rapidly as possible. The speed with which this is accomplished is apparent from the fact that one centrifuge handles the products of four nitrators, each of which produces a batch every 20 to 30 min. During the centrifuging process, the acid-wet cellulose nitrate absorbs moisture from the air. This is accompanied by some denitration with the development of heat. If the cellulose nitrate is not drowned quickly enough, the extent of denitration becomes great enough to be objectionable, and the cellulose nitrate may even decompose violently.

In recent years a continuous process for the nitration of cellulose has been developed. One of the main advantages of the process is that the drowning operation is avoided and more spent acid is recovered. This is accomplished by continuous, countercurrent washing of the crude, acid-wet cellulose nitrate.

Purification and stabilization are carried out in batches as large as 12,000 lb in tubs lined with stainless steel. The cellulose nitrate is first washed to a low level of acidity. The stabilization process includes one or more boiling treatments in very dilute mixed acid or pure water. The cellulose nitrate is water-washed between boils. Cellulose nitrate to be used in smokeless powder manufacture is reduced in fiber length in a Jordan refiner (used in paper manufacture) and then washed to remove traces of occluded acid. The product is finally steeped or boiled in dilute sodium carbonate and washed free of alkali.[7]

Nitro Compounds

Another class of compounds of wide and varied application results from substitution of a hydrogen atom in an organic compound with the nitro group, NO_2. These are typified by nitrobenzene, $C_6H_5NO_2$, the starting point for aniline, $C_6H_5NH_2$, (a crude sample of) from which Perkin produced the first coal-tar dye, mauve, in 1858; and by trinitrotoluene, TNT, already mentioned as a high explosive consumed in great quantities in both World Wars.

Nitration of hydrocarbons also requires (generally) the presence of sulfuric acid as well as nitric acid, usually in some excess, 10 to 20%. Here the role of the sulfuric acid is not merely that of removing water as formed but of providing a strong acid medium toward which nitric acid acts as a base and forms the nitronium ion:

$$HNO_3+H_2SO_4 = NO_2^+ + H_3O^+ + 2HSO_4.$$

[7] This description adapted, with permission, from R. E. Kirk and D. F. Othmer, Eds., *Encyclopedia of Chemical Technology*, 2nd Ed., Vol. 4, p. 625, Interscience, New York, 1964.

The nitration reaction then, is properly understood as following the equation:

$$RH + NO_2^+ + H_3O^+ + 2HSO_4^- = RNO_2 + H_3O^+ + H_2SO_4 + HSO_4^-.$$

The manufacturing process for nitrobenzene and for other mononitro compounds is not particularly complicated.

Modern nitrators hold approximately a 1000-gallon charge of benzene and operate batchwise on a 4-hour time cycle. The charge of benzene is transferred to an empty nitrator, the temperature adjusted, and the mixed-acid feed begun, typically containing about 41% HNO_3, 45% H_2SO_4, and 14% water.

The mixed acid is fed into the nitrator by gravity from a feed tank through a leg that introduces the acid either on the surface of the stirring mixture or just below the running agitator. The mole ratio of nitric acid to benzene is slightly less than 1.0 to keep dinitrobenzene formation at a minimum. The batch, which starts out at essentially room temperature, is brought up to the reaction temperature by the heat of reaction of the nitration process. For the manufacture of nitrobenzene this reaction temperature is between 45° and 60°C, the exact nitration temperature being a function of the mixed acid used, the design of the nitrator, and the specifications on the dinitrobenzene content of the final product. The time required for the mixed-acid feed is usually governed by the efficiency of the agitation, the heat-transfer surface available in the nitrator to remove the heat of reaction (625 Btu/lb benzene), and the cooling-water temperature and flow. After the mixed acid has been fed in, the batch is stirred for about one hour at the reaction temperature, or at a slightly higher temperature (70° to 90°C), in order to reduce the concentration of the residual nitric acid in the spent acid. The completed nitration is then transferred to the separator, where the batch is allowed to settle for 1 to 12 hours, depending on the particular design of the plant.

The spent acid is drawn off at the bottom of the separator and transferred to the spent-acid tanks for additional settling or to the spent-acid washers where the residual nitrobenzene and nitric acid are extracted from the spent acid with the next benzene charge.

The nitrobenzene layer in the separator is pumped to the neutralizers, which are vertical, lead-lined wooden tubs equipped with an air distributor for agitating the nitrobenzene during the washing process. The first wash given the oil is with the water from the final wash of the previous batch. The charge is thoroughly agitated and allowed to settle for 30 minutes before drawing off the supernatant liquor. This is followed by a neutralizing wash with warm sodium carbonate solution and a final water wash that becomes the first wash for the next batch. The losses occurring during neutralization are directly proportional to the number of washes and the quantities of water used. The amount of nitrobenzene actually lost in the wash waters is always slightly larger than the amount soluble, owing to entrainment losses.

If the washed nitrobenzene is to be used for the preparation of aniline, it is passed through a steam stripper to remove any unnitrated material. Nitrobenzene for other uses is generally refined by vacuum distillation of the washed crude and sold as "oil of mirbane." The distillation is carried out under reduced pressure in order to minimize the hazards that would be present at higher temperatures.[8]

The nitration of other aromatic compounds (derivatives of benzene) follows much the same pattern, modified as may be required, depending on the extent of nitration required, and the melting point of the products involved. Trinitrotoluene, TNT, desired as only the symmetrical, or α-isomer, 2,4,6-trinitrotoluene, is ordinarily made in three successive stages, the last being made at a relatively high temperature with a mixed acid containing an excess of SO_3 sulfur trioxide, made up from "fuming sulfuric acid" or "oleum."

Nitric acid is also employed in the manufacture of nitroparaffins, such as nitromethane, CH_3NO_2, and nitropropane $C_3H_7NO_2$. These find use as solvents and as intermediates for

[8] Adapted, with permission, from Kirk and Othmer, *Encyclopedia of Chemical Technology*, 1st Ed., Vol. 9, p. 389, Interscience, New York, 1952.

making other compounds. The manufacturing process does not involve mixed nitric and sulfuric acids, but is conducted in the vapor phase with nitric acid alone, at temperatures of 400° to 450°C. The yield of nitrocompound is not high, only around 40%. The remainder of the hydrocarbon reacting gives oxidation products, the nitric acid being reduced to nitric oxide that must be recovered.

Nitrogen peroxide can also be used as the nitrating agent. A yield of 50% of nitrocyclohexane has been reported, based on the cyclohexane reacting with NO_2 (some 16% of that supplied, in a cyclic process), the rest again going to oxidation products.

Oxidations Carried Out with Nitric Acid

Finally, nitric acid finds use as an oxidizing agent, as mentioned in Chapter Three. The application most widely used industrially is the example cited there, the oxidation to adipic acid, $HOOC{-}(CH_2)_4{-}COOH$, one of the two intermediates needed in the manufacture of the most popular kind of nylon, the 6—6 variety (the other ingredient being hexamethylene diamine), by the oxidation of a mixture of cyclohexanol and cyclohexanone. This mixture is made by catalytic air oxidation of cyclohexane derived from petroleum.

The process has been described only briefly in the literature. The mixture of intermediate oxidation products is fed to a stainless steel reactor, provided with means for cooling, where it is oxidized at about 180°F (85°C) with nitric acid of about 60% strength. Small quantities of copper and vanadium salts serve as catalysts. The equation given on p. 52 of Chapter Two for this oxidation reaction showed nitrogen dioxide as the only reduction product of the nitric acid. Actually, it is found that much, maybe three-fourths, of the nitrogen in the nitric acid reacting goes to nitrous oxide or to elemental nitrogen,

$$C_6H_{11}OH + 2HNO_3 = (CH_2)_4{-}(COOH)_2 + N_2O + 2H_2O,$$

and is not recoverable. The nitrogen dioxide that is formed is recovered by the usual procedure of oxidizing with air and countercurrent absorption in water. The adipic acid, mp 152°C, is obtained by crystallization and, upon cooling, centrifugally separated and purified by recrystallization.

A similar process has been used to produced oxalic acid, the simplest dibasic acid, $(COOH)_2$, from glucose or materials largely made up of glucose (hydrolyzed starch or oat hulls or molasses):

$$C_6H_2O_6 + 6HNO_3 = 3(COOH)_2 + 6NO + 6H_2O.$$

The process has, however, been largely displaced by one starting with sodium formate, which can be decomposed to sodium oxalate and hydrogen on heating.

These and other processes which have been proposed from time to time seem to have this feature in common: Nitric acid as an oxidizing agent is best employed in the preparation of useful products when it is applied to starting materials already partially oxidized, not to hydrocarbons directly.

In all these reactions I hope you will recognize the inherent possibility of "runaway" reactions. If the temperature gets out of control, you have present a strong oxidizing agent together with oxidizable material, and there is a marked tendency for the reaction to run to completion, producing carbon dioxide and water and elementary nitrogen, with an even greater evolution of heat.

It is this property of nitric acid that brings it into consideration as an oxidizer in rocket propellant combinations. You do not hear as much about it as you do about liquid oxygen, but it does find important specialized applications.

Rocket propellants, as you may have read elsewhere, are rated on the basis of their "specific impulse," I_{sp}, often given (incorrectly) in seconds; more exactly, it is a measure of the pounds of thrust exerted by the combustion and expulsion through a nozzle of 1 lb/sec of the propellant or propellant mixture. To give an idea of the values involved, the combination most used, the same type as used in the German "V-2," of liquid oxygen and kerosine ("RP-1") gives $I_{sp} = 265$. Liquid oxygen plus liquid hydrogen, more recently becoming available, gives $I_{sp} = 365$.

Liquid oxygen is fine for launching research vehicles, where time is available for loading and reloading if the shot has to be postponed. That would never do for a military missile; it has to be ready to blast off instantly. This requirement has tended to favor solid propellants. Two of these have been mentioned incidentally already: double-base explosives (nitroglycerin plus nitrocotton), and composite propellants made from ammonium perchlorate, plus aluminum powder, and a binder. These both give I_{sp} of about 250.

Solid propellants are not as controllable as liquid, and military requirements could be met by using storable liquid components. This is where nitric acid comes into the picture. High-strength, pure nitric acid, sometimes called, ineptly, "white fuming nitric acid," WFNA, is rather difficult to store safely, even in aluminum. It does tend, as you know, to decompose. "Red fuming nitric acid," RFNA, containing about 6% NO_2 in solution, would be stable but is rather corrosive. The addition of about 15% of sulfuric acid makes it possible to store the mixture, IRFNA, in steel, though some sludge is formed that has to be taken care of before firing the rocket.

Storage is actually easier with liquid nitrogen peroxide, as we have been calling it; it boils at 21.15°C, and a container would have to hold it at a moderate pressure.

The third stage of the "Atlas Agena" rocket is powered with IRFNA, and N_2O_4 is the oxidizer for the Air Force "Titan II" missile. The fuel in the Agena is a derivative of ammonia, unsymmetrical dimethylhydrazine, UDMH, $H_2N—N(CH_3)_2$, giving I_{sp} of 275. In the Titan II the fuel is a mixture of hydrazine, $H_2H—NH_2$, with UDMH, giving I_{sp} of 285. These (and some other) fuels with either "fuming" nitric acid or N_2O_4 give propellants that are called "hypergolic," indicating that they are self-igniting when brought together.

If you watched the Gemini V flight in 1965, you may have heard them say that the "thrusters" on board, for changing the speed and direction of the space capsule, were powered with nitrogen tetroxide and hydrazine.

The plans for the Apollo Mission to the moon provide for the use of this same combination, or one very similar, to power the "service module" of the spacecraft on its take-off for the return journey to earth.

That brings us now to the dawn of tomorrow, to the brink of outer space. I can think of no better way to close this little book devoted to the process which Wilhelm Ostwald pioneered than to quote a maxim attributed to him:

> *Was war, hat seine Zeit gehabt.*
> *Was werden will, verdient unsere ganze Hingabe.*
>
> What was has had its day.
> What will be deserves our complete devotion.

APPENDIX A

Nitric Acid Production in the United States

Thousands of tons per year, as 100% HNO_3	
1929	34
1934	—
1939	168
1944	472
1949	1189
1954	2289
1959	3074
1964	4609

Source: U.S. Bureau of the Census.

APPENDIX B[1]

Potassium Nitrate, KNO_3, from Sodium Nitrate and Potassium Chloride

When sodium nitrate and potassium chloride are dissolved, the solution contains four ions, Na^+, NO_3^-, K^+, Cl^-, and from these ions not only could the two original salts be reconstructed, but also two new salts, potassium nitrate and sodium chloride, through a regrouping of the radicals. Which of the four salts will crystallize from a solution containing the four ions depends solely on their solubilities, which may vary much or little with the temperature. The following table gives the solubility at different temperatures for each salt. Thus, at 10°, 21 grams of KNO_3 are soluble in 100 grams of water. This means that, if an excess of solid potassium nitrate is shaken with pure water until no more will dissolve, the clear solution will then contain 21 grams of

Grams of Salt Soluble in 100 Grams of Water

	At 10°C	At 100°C
KNO_3	21	246
NaCl	36	40
KCl	31	56
$NaNO_3$	81	180

[1]Adapted from A. A. Blanchard, J. W. Phelan, and A. R. Davis, *Synthetic Inorganic Chemistry*, Chap. II, pp. 53–58, 5th Ed., Wiley, New York, 1936, with permission.

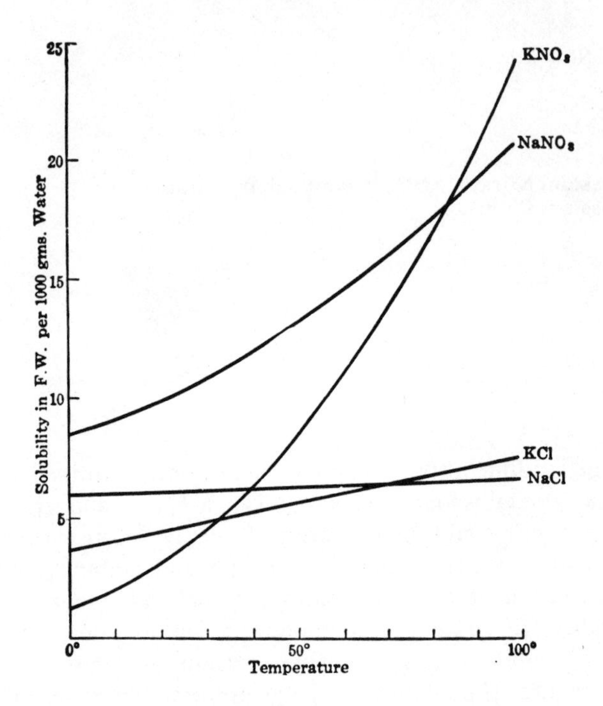

Figure B.1 Variation of Solubility with temperature.

From A. A. Blanchard, J. W. Phelan and A. R. Davis, *Synthetic Inorganic Chemistry*, Chap. II, pp. 53–58, 5th ed., Wiley, New York, 1936, with permission.

KNO_3 for every 100 grams of water. On the other hand, if a solution of 42 grams of KNO_3 in 100 grams of water obtained at a higher temperature is cooled to 10° and stirred until an equilibrium is attained, all but 21 grams of the salt crystallizes out, and the solution has exactly the same concentration as that obtained in the other way.

A very important fact concerning solubilities is that the solubility of a given salt is practically unaffected by the presence of another salt in the solution, provided only that the other salt does not possess one of the same ions as the first salt.

For example, suppose that sodium nitrate and potassium chloride in equivalent amounts are added to 100 grams of water at 10°, so that the total weight of K^+ and NO_3^- radicals will be 42 grams; the potassium nitrate in excess of its solubility will then crystallize out, and 21 grams of the crystals will thus be obtained. The presence of the radicals of sodium chloride in the solution is without effect on the potassium nitrate.

In the following procedure 2 *formula weights* (F.W.) of $NaNO_3$ (170 grams) and 2 F.W. of KCl (149 grams) are together treated with 210 grams of water at the boiling temperature. The 170 grams of $NaNO_3$ will dissolve in the water since this amount will not nearly saturate 210 grams of water, but from the table it is seen that it will take only 118 grams of KCl to saturate this amount of water. If this were the only consideration we should expect 118 grams to dissolve and 31 grams of KCl to remain undissolved. But we must consider the presence of the Na^+ ions also in conjunction with the Cl^- ions, and since only 84 grams of NaCl are soluble in 210 cc of water and in all 2 F.W. or 117 grams of NaCl are available we may conclude that 33 grams will crystallize out. This removes Cl^- from the solution and upsets the equilibrium which would otherwise exist between the solid and dissolved KCl. Thus the entire KCl will dissolve and furnish the entire 2 F.W. of Cl^- ions for the 2 F.W. of NaCl. The K^+ and NO_3^- ions thus made available constitute 2 F.W. or 202 grams of KNO_3, which according to the table would be completely soluble in 210 cc of water at 100°. Hot filtration at this point will retain a part of the NaCl as solid in the filter but allow the whole of the 2 F.W. of KNO_3 to pass into the filtrate. Cooling of the filtrate will allow a new state of equilibrium to be established corresponding to the solubilities at the lower temperature.

FLOW SHEET

Heat in a covered casserole 170 grams of $NaNO_3$, 149 grams of KCl, and 210 cc of water. Boil the mixture 1 minute and then filter hot. Do not rinse out the casserole but use it for the second boiling.

<table>
<tr><td>On Filter:</td><td colspan="4">Filtrate (B): Cool to 10° and filter.</td></tr>
<tr><td rowspan="3">NaCl, dirt and some KNO_3. Transfer to casserole in which first boiling was made.</td><td rowspan="3">On Filter:
Crystals of KNO_3 impure. Press and wash with 20 cc ice-cold water.</td><td colspan="3">Filtrate (D): Saturated with KNO_3 and NaCl. Pour it into original casserole containing impure NaCl (A). Boil 5 minutes. Filter hot.</td></tr>
<tr><td rowspan="2">On Filter:
NaCl dirt and a little KNO_3.
Save temporarily.</td><td colspan="2">Filtrate: Cool to 10° and filter.</td></tr>
<tr><td>On Filter:
KNO_3 impure. Press and wash with ice-cold water.</td><td>Filtrate: Saturated with KNO_3 and NaCl. Save temporarily in flask labeled "mother liquors."</td></tr>
<tr><td>(A)</td><td>(C)</td><td>(E)</td><td>(F)</td><td>(G)</td></tr>
</table>

RECRYSTALLIZATION

Unite the two lots of moist impure KNO_3, add half their weight of distilled water, and heat until solution is complete. Cool to 10° and filter, pressing out as much as possible of the liquid. Stop suction. Pour 15cc ice-cold distilled water over the crystals and let it permeate the mass. Apply suction and pressure. Test for chloride. If any is found repeat the washing process until the product is free from chloride. Add all filtrates to the mother liquors, *G*.

Materials: crude Chile saltpeter, $NaNO_3$, 170 grams = 2 F.W.
crude potassium chloride, KCl, 149 grams = 2 F.W.

Reagent: $AgNO_3$ solution.

Apparatus: 750-cc casserole.
5-inch watch glass.
5-inch funnel.
perforated filter plate.
suction flask and trap bottle.
250°C thermometer.
iron ring and ring stand.
Bunsen burner.

Procedure: Place the $NaNO_3$ and KCl in a 750-cc casserole. Add 210 cc of water, cover with a watch glass, and place over a low flame. While keeping watch of the casserole to see that its contents do not boil, prepare a suction filter. Then raise the flame under the casserole and watch it until boiling commences. Lower the flame, and let the mixture boil gently just 1 minute, keeping the watch glass over the casserole to retard evaporation. While it is at the boiling temperature pour the mixture from the casserole into the suction filter. Quickly scrape most of the damp salt into the filter, and cover the funnel with a watch glass to retard cooling. Do not rinse out the casserole but reserve it with the adhering salt and to it add the batch of crystals (*A*) as soon as they have been drained from the hot liquor. Pour the filtrate (*B*) into a beaker, and cool it to 10° by setting it in a pan of cold water or ice water.

Stir the crystallizing solution so as to obtain a uniform crystal meal which is easier to handle and drain on the filter than the larger crystals that would be obtained by slow cooling. Separate the KNO_3 crystals (*C*) from the cold liquor by means of the suction filter. Press the mass of crystals in the filter to squeeze out as much of the solution as possible. The thin layer of solution left adhering to the surfaces of the crystals is saturated with NaCl. To remove the greater part of this, stop the suction, pour carefully over the surface of the crystals 20 cc of ice-cold distilled water, let this water penetrate into the mass of crystals for perhaps 30 seconds, then apply suction and drain these washings into the rest of the filtrate. Pour the filtrate (*D*) into the casserole in which the batch of NaCl crystals (*A*) is reserved. Bring this mixture to the boiling point and boil gently 5 minutes with the casserole uncovered, thus allowing a part of the water to escape by evaporation. Then filter at the boiling temperature exactly as in the first instance. Cool the filtrate, and collect and rinse a second crop of KNO_3 crystals (*F*) in exactly the same manner as the first crop (*C*). Reserve the filtrate in a flask labeled "mother liquors" (*G*). Add together the two crops of KNO_3 crystals (*C* and *F*), and test them for chloride by dissolving about 0.1 gram in 2 cc of water and adding 5 drops of dilute HNO_3 acid and 1 drop of $AgNO_3$ solution. A considerable white precipitate will be seen, indicating the presence of chloride, and the crystals must be purified by recrystallization. Weigh the crystals while they are still moist, add to them in a beaker one-half this weight of distilled water, and warm until solution is complete.[2]

Cool the solution to 10° with stirring, and pour the mixture on to the suction filter. Some of the filtrate, which is saturated with KNO_3, may be poured back into the beaker to rinse the last of the crystals on to the filter. Drain the crystals with suction, pressing them down compactly in order to squeeze out as much of the solution as possible. Stop the suction. Pour carefully over the crystals 15 cc of ice-cold distilled water. Let it permeate the whole mass, then again apply suction and drain and collect all the filtrate and washings in the flask of "mother liquors" (*G*). Again test for presence of chloride. A trace will probably be shown. Continue the washing process, using 15-cc portions of ice water each time until no test is shown for chloride. Transfer the crystals to white paper towels. Fold the towels over the crystals to make a compact package, and leave the package over night to dry at room temperature. Transfer the dry crystal meal to a dry 4-ounce bottle and label the preparation neatly.

If a sufficient quantity of pure product is not obtained, all the mother liquor should be boiled down in the 750-cc casserole and used as a starting point in repeating the foregoing procedure.

Fifty grams may be regarded as a satisfactory yield.

[2]If the solution is not perfectly clear at this point it must be filtered. Dilute it with 50 cc of water so that it will not "freeze" on the filter. Pour it at the boiling temperature without suction through a filter, and then pour 20 cc of boiling water around the upper edge of the filter, letting it run through into the filtrate to carry the last of the KNO_3. The solution will have to be evaporated to bring it back to its original volume.

APPENDIX C

PROFITABILITY ESTIMATE FOR THE PRODUCTION OF POTASSIUM NITRATE

You are asked to estimate the investment that could be justified if a 30% return is expected annually before taxes (15% after taxes) for the production of potassium nitrate at two rates: 1 ton (2000 lb) per day, and 10 tons/day. Operation 24 hr/day, 360 days/year.

Raw materials: Potassium chloride (97% KCl) $28.30/ton
Sodium nitrate (95% $NaNO_3$) $44.00/ton
Products: Potassium nitrate (99% KNO_3) $190.00/ton
Sodium chloride—negligible value.
Labor requirement: two men per shift, independent of scale of operation; wage rate $2.50/hr. Overhead at 50% of wages.
Steam, cooling water, electricity, miscellaneous supplies: estimate at $2.50/ton KNO_3.
Repairs, maintenance, other items related to investment: estimate at 10% annually on investment.
Depreciation (amortization set-aside): take at 10% on investment, annually.

Note that the several items related to investment total 50%/year on the investment. It follows that the justifiable investment, under the conditions specified, is $(\frac{1}{0.50}) = 2.0$ times the net difference between what can be expected from sales and the out-of-pocket expenses per year; that is, raw materials, labor (plus overhead), utilities, and supplies. Assume that a yield can be obtained of 90% of theoretical.

APPENDIX D

PREPARATION OF NITRIC ACID FROM SODIUM NITRATE AND SULFURIC ACID[1]

Take a 250-cc retort with a glass stopper, and set up as shown in the diagram. The test tube which is to serve as a receiver is to rest in a dish or large beaker containing cold water, and should be covered with a wet cloth. In the retort place 20 grams of $NaNO_3$ and 20 cc of concentrated H_2SO_4. Heat until most of the HNO_3 has been driven over, leaving a nearly solid residue $NaHSO_4$, in the retort.

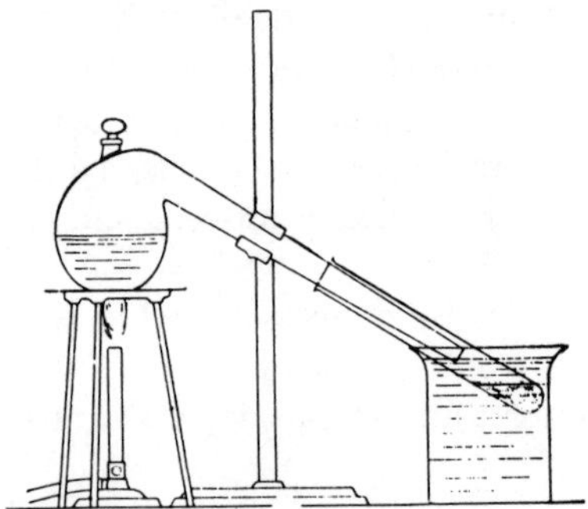

Figure D.1 Retort for preparation of nitric acid.

From H. G. Deming and S. B. Aronson, *Exercises in General Chemistry and Qualitative Analysis*, 3rd ed., Wiley, New York, 1930, with permission.

[1] Adapted from H. G. Deming and S. B. Aronson, *Exercises in General Chemistry and Qualitative Analysis*, 3rd Ed., Wiley, New York, 1930, with permission.

To avoid breaking the retort, proceed as follows: Add warm (not hot) water and allow to soak for 15 minutes. Then carefully shake to dissolve the solid. If one such treatment is not sufficient, repeat.

Note: it is best to neutralize acid products and residues with alkali (sodium bicarbonate will do) and dilute them liberally with water before disposing of them to the drain, especially in a home laboratory.

It would be well to have a pan under the apparatus sufficient to hold all the reagents in case of breakage. Caution: You should not do any laboratory experiments without eye protection.

INDEX